W9-CEH-529

ALICE MAH

Industrial Ruination, Community, and Place

Landscapes and Legacies of Urban Decline

UNIVERSITY OF TORONTO PRESS
Toronto Buffalo London

Toronto Buffalo London
www.utppublishing.com
Printed in Canada

ISBN 978-1-4426-4549-3 (cloth)
ISBN 978-1-4426-1357-7 (paper)

Printed on acid-free, 100% post-consumer recycled paper with vegetable-based inks.

Library and Archives Canada Cataloguing in Publication

Mah, Alice, A. (Alice Anastasia), 1978–
Industrial ruination, community, and place: landscapes and legacies of urban decline / Alice Mah.

Includes bibliographical references and index.
ISBN 978-1-4426-4549-3 (bound). ISBN 978-1-4426-1357-7 (pbk.)

1. Shrinking cities – Case studies. 2. Shrinking cities – Social aspects – Case studies. 3. Collective memory – Case studies. 4. Deindustrialization – Social aspects. 5. Collective memory and city planning. I. Title.

HT151.M32 2012 307.76 C2012-902649-2

This book has been published with the help of a grant from the Canadian Federation for the Humanities and Social Sciences, through the Aid to Scholarly Publications Program, using funds provided by the Social Sciences and Humanities Research Council of Canada.

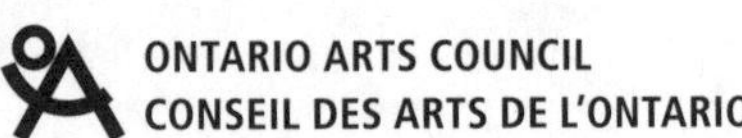

University of Toronto Press acknowledges the financial assistance to its publishing program of the Canada Council for the Arts and the Ontario Arts Council.

University of Toronto Press acknowledges the financial support for its publishing activities of the Government of Canada through the Canada Book Fund.

Contents

Figures

Acknowledgments

This research was funded by generous grants from the Social Sciences and Humanities Research Council of Canada, the Dr Ingeborg Paulus Scholarship, and the University of London Central Research Fund. The writing of this book was also made possible by a Research Fellowship at the International Research Center for Work and Human Life Cycle in Global History at Humboldt University in Berlin (2009–10).

Parts of this book have been previously published in different forms in academic journals. An earlier version of chapter 3 was published as: Mah, A. 2010. "Memory, Uncertainty and Industrial Ruination: Walker Riverside, Newcastle upon Tyne." *International Journal of Urban and Regional Research*, 34 (2): 398–413. A previous version of chapter 6 was published as: Mah, A. 2009. "Devastation but also Home: Place Attachment in Areas of Industrial Decline." *Home Cultures*, 6 (3): 287–310.

I would like to thank the following individuals for their guidance, help, and encouragement: Fran Tonkiss, Richard Sennett, Noel Whiteside, Graham Crow, Andy Pratt, Ash Amin, Wally Clement, Rianne Mahon, Jennifer Cottrill, Jacob Eyferth, Eileen Church Riley, and Roger Stephens. Many thanks are due to Doug Hildebrand and Virgil Duff at the University of Toronto Press for their enthusiasm and support for this book. I would also like to thank the anonymous reviewers of the manuscript for their constructive criticisms and comments. Special thanks are due to all of the people who I interviewed in Niagara Falls, Newcastle upon Tyne, and Ivanovo. Finally, I would like to thank my parents, Eric and Kathy Mah, and my partner, Colin Stephen, for their tremendous patience, love, and support.

INDUSTRIAL RUINATION,
COMMUNITY, AND PLACE

Chapter One

Introduction

Industrialized cities around the world feature derelict factories, mills, warehouses, and refineries. Once behemoth structures at the social and economic heart of industrialization, these buildings now lie in ruins. The scale of this decay echoes the grandeur of fallen past civilizations, yet industrial ruins were produced within a much shorter time frame. Abandoned industrial buildings have captured the aesthetic and sociological imagination of scholars, travellers, artists, and journalists. Across the globe, they have been occupied by artists, musicians, and squatters; appropriated for cultural or consumption uses; and photographed, painted, and documented. Yet old industrial sites are invested with more than cultural meanings: they are the remnants left behind in the wake of deindustrialization. Despite their state of disuse, abandoned industrial sites remain connected with the urban fabric that surrounds them: with communities; with collective memory; and with people's health, livelihoods, and stories.

Industrial ruins are alternatively left abandoned, re-used, regenerated, sold, or demolished. They are never static objects, but are in a constant state of change across time and space. Thus, this study of industrial ruins is framed in terms of "ruination" rather than "ruins," because the word "ruination" captures a process as well as a form. This book is concerned with the relationship between industrial ruination, community, and place, specifically, the *landscapes* (socio-economic and cultural geographies) and *legacies* (the long-term socio-economic and psychological implications for people and places) of the interrelated processes of industrial ruination and urban decline. I will explore these complex relationships and processes through three paradigmatic case studies in Niagara Falls, Canada / USA; Newcastle upon Tyne, United Kingdom, and Ivanovo, Russia. The case studies are of old industrial communities in different national and political contexts that have experienced significant deindustrialization in the latter half of the twentieth century.

Deindustrialization in the 1970s and 1980s in North America and Western Europe attracted considerable academic attention across a range of disciplines, with debates on the role of the state versus the market, the importance of manufacturing versus services, and the antithesis between "community" and "capital" (see Alderson 1999; Bluestone and Harrison 1982; Cowie and Heathcott 2003; High 2007; High 2003; Rowthorn and Ramaswamy 1997; Staudohar and Brown 1987). In *Capital Moves*, Jefferson Cowie (1999) vividly illustrates the devastating social costs of deindustrialization for communities by following one company as it pursues cheap and flexible labour from Camden, New Jersey, to Bloomington, Indiana, in the 1940s, to Memphis, Tennessee, in the 1960s, and finally to Ciudad Juarez, Mexico. In a similar vein, Bluestone and Harrison (1982) describe the negative impact of economic restructuring on income distribution in the United States as representing an "hourglass economy" comprising many high-skilled information technology and knowledge sector jobs that generate high income, few jobs that generate middle income, and many low-skilled service jobs that generate low income. By contrast, other scholars locate deindustrialization within "inevitable" processes of economic change in a market economy, downplaying its negative impacts on people and places (cf. Alderson 1999; Rowthorn and Ramaswamy 1997).

The research discussed in this book highlights the impacts of deindustrialization on place-based communities (as contrasted with communities based on interest or identity), and focuses on residential areas adjacent to sites of industrial ruination. Community studies of place have a long and varied history of scholarship, including a number of classic studies of disadvantaged communities in particular (see Bell and Newby 1971; Lassiter et al. 2005; Mumford and Power 2003; Stacey 1969; Winson and Leach 2002; Young and Willmott 1957). As Crow (2002, 3.2) argues, community studies have the potential to ground, test, and challenge abstract theories of social change, "allowing researchers to explore what processes like globalization and deindustrialization actually mean for the everyday lives of ordinary people at a local level." However, many authors have criticized the term "community" because of its relationship to romanticized and nostalgic notions of social cohesion and its tendency to represent neighbourhoods as "relatively class-homogeneous, small-scale, easily delineated areas with clear borders, hosting relatively cohesive communities" (Blokland 2001, 268). While I use the concept of community within this work, I recognize that it is a contested term.

Some studies have gone beyond the politicized antithesis between capital and community to examine the shifting values, expectations, and lived experiences of deindustrialization and post-industrial change. For example, in *Industrial Sunset: The Making of North America's Rust Belt, 1969–1984*, Steven High (2003) draws on a wide range of sources, including songs, poetry, archival material, and oral history interviews with displaced workers, to explore the complex meanings of job loss in Canada and the United States. High (2003, 17) argues that Canadian workers were more successful than American workers in softening the blow of job losses through adopting an ideology of Canadian economic nationalism, which prevented both the image and reality of the Rust Belt from entering the country. In another study, Ruth Milkman (1997) explores the social and cultural impacts of job losses on factory workers at a GM plant in New Jersey. Contrary to many accounts of deindustrialization, which portray workers as nostalgic about factory work, she shows that many workers in the GM plant welcomed the possibility of change because they disliked the old factory system, especially abusive and degrading treatment by supervisors. Similarly, Kathryn Dudley's (1994) anthropological study of deindustrialization and transition to an uncertain "post-industrial" future in an American Rust Belt company town highlights complexities in the lived experiences of blue-collar workers. Dudley argues that the American dream, based on the belief that anyone who is willing to work hard is able to be successful, often proves false under conditions of economic contraction. This book follows High in its concern with comparing Canadian and American experiences of deindustrialization within the Rust Belt, and follows both Milkman and Dudley in its concern with cultural meanings and lived experiences of deindustrialization. My analysis also broadens the scope of study to encompass different industries and geographies, and remains anchored in economic as well as political, social, and cultural dimensions of industrial ruination.

The shift in advanced industrial countries from manufacturing to services has been theorized in various ways as a qualitative shift to a new type of economy and society. My own analysis is situated within the context of a key socio-economic shift or transformation, whether it is framed in terms of a shift from manufacturing to services, from the industrial to the post-industrial, from Fordism to post-Fordism, from the old economy to the new economy, from modernity to postmodernity, or from the international to the global (Beck 1992; Bell 1973; Boyer and Durand 1997; Giddens 1991; Harvey 1989; Jessop 1991; Leadbetter 1998; Lipietz 1992). Of course, the extent to which any one of these processes of change is accurate or complete is debatable, and many theorists would concur that elements of "older" structures coexist within the new. Perhaps more controversial than the scale or extent of these changes is the value attached to them: some perceive the new, flexible economy as a space for greater opportunity, social mobility, and advancement, while others argue it is characterized by greater risk, socio-economic polarization, uncertainty, and instability. Some lament the demise of older social and economic arrangements, while others celebrate the change or accept it as inevitable. By focusing on the people and places that have been "left behind" in the new economy, my stance is more critical of the broad implications and assumptions of post-industrial transformation.

Research on deindustrialization and economic restructuring has primarily concentrated on the industrialized West, yet these trends are increasingly visible throughout the world.[1] By focusing on case studies in different national contexts, this book explores both global and local dimensions of industrial decline. While most studies of deindustrialization focus on the immediate impacts of plant closures on labour, capital, and communities, this research analyses the same processes through a broader historical, geographical, and theoretical lens. Borrowing insights from "contemporary archaeologies of the recent past" (Buchli and Lucas 2001), this book aims to "read" the past within the present in order to better understand the present, and to better explore the long-term material, social, and psychological implications of industrial decline for people and places around the world.[2]

Methodologically, this research is guided by the notion that the study of waste, of what is discarded, is sociologically important. My concern is with landscapes of industrial ruination and urban decline as "wasted places" that have, for various reasons, yet to be transformed. There are many stories of "winners" in the literature on cities and regeneration, following the model of arts-led regeneration exemplified by the gentrification and urban redevelopment of SoHo in Manhattan in the 1980s (Zukin 1982). According to O'Connor (1998, 229), this model of urban regeneration "was based on a conscious and explicit shift of the economic base from manufacturing to services industries, symbolized by the redrawing of the old historical industrial areas in terms of leisure and consumption." Recently, many old industrial cities have adopted the "creative cities" approach, which involves coordinated urban attempts at arts- and culture-led regeneration and branding (Florida 2005; Landry and Bianchini 1995). Ironically, many cities copy creative models; for example, through the fashionable conversion of old industrial buildings into museums, art galleries, studios, or lofts. With the increasing number of cities adopting this approach, creative cities are becoming competitive (Doel and Hubbard 2002; Hall and Hubbard 1998) as each vies for government funds, corporate investment, and to become a symbol of arts- and culture-led regeneration. One city that has been widely recognized as a success in regeneration is Manchester, which consciously adopted a place-marketing strategy in the 1990s that commentators termed the "entrepreneurial city" or "competitive city" approach (Quilley 1999; Ward 2003).

But not all cities can succeed in a competitive model. There are many stories of those that do not – of cities stigmatized by social and economic deprivation, poor infrastructure and public services, dilapidated housing, depopulation, and unemployment. These stories tend to be overlooked in the interest of a progress-oriented view – of moving on in the capitalist process of "creative destruction" (Schumpeter 1965). Schumpeter argues that in order for capitalism to remain dynamic and innovative, it must undergo both creation and destruction:

> The opening up of new markets, foreign or domestic, and the organizational development from the craft shop and factory to such concerns as U.S. Steel illustrate the same process of industrial mutation – if I may use that biological term – that incessantly revolutionizes the economic structure from within, incessantly destroying the old one, incessantly creating a new one. This process of Creative Destruction is the essential fact about capitalism. It is what capitalism consists in and what every capitalist concern has got to live in. (Schumpeter 1965, 83)

The concept of creative destruction helps to explain economic processes of deindustrialization and economic decline within capitalist development, but it also informs the market-based rationale of urban development and regeneration policies. This economic perspective focuses on growth, innovation, and selective renewal, rather than dwelling on the "inevitable" waste left behind.

My research focuses on the people and places that have been left behind within the context of an uneven geography of capitalist development. According to several critical geographers (Harvey 1999; Massey 1984;

Smith and Harvey 2008), capitalism produces an inherently uneven geography of development whereby industries, people, and places are constantly abandoned in capital's search for new sites and cheaper inputs (a geographical manifestation of creative destruction). In *The Limits to Capital*, Harvey (1999) argues that capital has inherent tendencies towards concentration, crowding, and agglomeration, and thus encounters physical, social, and spatial limits. The tendency for capitalism to produce crises of over-accumulation, or surpluses of capital and/or labour, is periodically offset or absorbed by a "spatio-temporal fix": temporal displacement through investment in long-term capital projects or social expenditures, or spatial displacements through opening up new markets or production capacities elsewhere. The geographical relocation of industries to places with cheaper inputs (absorbing surpluses of capital and labour) is an example of a spatio-temporal fix, but so too are efforts to regenerate old industrial areas with new development centred around services, real estate, and finance.

Harvey's work is an important theoretical starting point for investigating the uneven and contradictory ways in which capitalism produces and reproduces landscapes of industrial ruination. Similarly, this research draws on Zukin's (1991, 5) argument that post-industrial places are "sharply divided between landscapes of consumption and devastation," and that landscape is the cultural product of institutions of power, class, and social reproduction. Although my research was inspired partly by the theories of both Harvey and Zukin, I found that the distinctiveness and complexity of landscapes and legacies of industrial ruination cannot be accounted for by the binaries of success and failure, creation and destruction, or consumption and devastation. There are many tensions, contradictions, and contingencies within the lived experiences of people who occupy these "wasted places." As Doreen Massey (1984, 299–300) argues:

> Capitalist society, it is well-recognized, develops unevenly. The implications are twofold. It is necessary to unearth the common processes, the dynamic of capitalist society, beneath the unevenness, but it is also necessary to recognize, analyse and understand the complexity of the unevenness. Spatial differentiation, geographical variety, is not just an outcome: it is integral to the reproduction of society and its dominant social relations. The challenge is to hold the two sides together; to understand the general underlying causes while at the same time recognizing and appreciating the importance of the specific and the unique.

In other words, the role of capital in shaping uneven geographies does not always conform to a straightforward pattern of capital flight, massive job losses, and burdened communities: it is important to look at both the general and the unique to understand these processes. This book seriously considers Massey's insistence on recognizing the complexity of the unevenness of capitalist development. Each case study in this book reveals complex social, cultural, and economic dynamics that relate both to wider processes of uneven development and to local dynamics.

This book advances a new theoretical framework for analysing the complex relationships between deindustrialization and industrial ruins: "industrial ruination as a lived process." Working within this framework reveals two related insights. First, industrial ruins are not simply forms, but rather they are embedded within processes of creation and ruination. Each phase of industrial ruination is situated along a continuum between creation and destruction, fixity and motion, expansion and contraction. Over time, landscapes of industrial ruination will become landscapes of regeneration, reuse, demolition, or ruination once again. Second, forms and processes of industrial ruination are experienced by people. Many people experience industrial ruins indirectly, from a distance: from the window of a moving car, bus, or train; during a visit to an unfamiliar city, neighbourhood, or stretch of road; though the lens of a camera; or as an act of tourism through deliberately seeking out ruins as sites for art, play, or mischief. However, as this book will show, many people also experience industrial ruins more directly, from inside rather than from outside: they live in and among industrial ruins and identify them as home.

Industrial Ruination

The concept of ruins implies finality, beauty, majesty, glorious memory, tragedy, loss, and historical import. According to Jakle and Wilson (1992), ruins reflect the past, romance, and nostalgia, and at the same time represent risk, commodification, and neglect. The authors hit upon a crucial point about the spatially uneven violence of capitalism in the management and treatment of ruins: certain buildings are maintained or renewed, while others are left to fall into disrepair. This has more to do with patterns of capitalist accumulation and expansion than it does with any natural structural life-cycle. Similarly, Walter Benjamin (2000, 13) wrote of the "ruins of the bourgeoisie" as the necessary but unfortunate outcome of the progress of history, modernity, and capitalism.

Industrial ruins are produced by capital abandonment of sites of industrial production; the sites that are no longer profitable and that no longer have use-value can be read as the footprint of capitalism. A number of authors have described derelict landscapes as wasted cultural, social, and economic spaces (cf. Berman 1983; Cowie and Heathcott 2003; Harvey 2000; Jakle and Wilson 1992; Stewart 1996; Zukin 1991). However, industrial ruins as material sites of investigation have been studied primarily in relation to art, photography, and culture (Edensor 2005; MacKenzie 2001; Stewart 1996; Vergara 1999).

Aesthetic and cultural studies of industrial ruins simultaneously mourn and celebrate the landscape of industrial ruins, on the one hand as sad beauty, and on the other hand as the genuine cultural ruins of civilizations. The celebration of industrial ruins is most obvious within dereliction tourism, represented by a number of websites devoted to virtual tours of derelict buildings across the globe. Three notable examples of dereliction tourism are Paul Talling's "Derelict London" website, which is devoted to derelict spaces in London and has over 800 photographs with subjects such as disused railway lines, cemeteries, shops, pubs, waterways, and public toilets (www.derelictlondon.com); Lowell Boileau's "Fabulous Ruins of Detroit" website, which displays photos of decrepit buildings as evidence of the glorious past of the American industrial age that is meant to parallel the great ruins of Europe, Africa, and Asia (www.detroityes.com); and Uryevich's "Abandoned" website, which highlights the aesthetic, cultural, and spiritual importance of "abandoned things" in the former Soviet Union and includes 559 photographs of abandoned buildings, plants, and industrial sites (www.abandoned.ru). The book *Industrial Ruins* (Edensor 2005) is another celebratory example, which encourages people to enjoy industrial ruins as spaces of leisure and imagination and criticizes the notion that industrial ruins are wasted spaces. There are more mournful and ambivalent examples in photojournalist studies of urban and industrial ruins, including *Manufactured Landscapes: The Photographs of Edward Burtynsky* (Pauli et al. 2003) and *American Ruins* (Vergara 1999), with stunning photographs accompanied by social commentary on the implications of industrial ruins for society and the environment.

My research of industrial ruination as a lived process brought me towards a criticism of artistic and cultural studies of industrial ruins, particularly studies that ignore the lives of the inhabitants of ruined landscapes. Nonetheless, the rise of dereliction tourism and artistic interest in abandoned industrial sites is sociologically interesting because it shows that these sites are formidable enough in their presence to capture the popular and artistic imagination. The staggering social, economic, political, and geographical impacts of industrialization spurred profound sociological and artistic inspiration through the writings of Dickens and Engels and the paintings of Turner and the Futurists, to name a few examples. It is not surprising that the impacts of deindustrialization should spark their own sociological and artistic response. This book was also inspired by the visual impact of industrial ruins, but my interest is in the complex stories that stretch beyond mourning and celebration.

To view something as a ruin is already to have a perspective. Urban planners see industrial ruins as potential sites for redevelopment into museums, art galleries, or trendy apartments. Political economists see industrial ruins as the waste products of an uneven geography of capitalist development. Dereliction tourists, artists, and photojournalists see industrial ruins as beautiful yet tragic physical reminders of mortality and finality. But where some people see ruins, others see homes situated within painful processes of transformation. Rather than framing industrial ruins as fixed forms with implicit meanings related to economic or aesthetic value, my analysis focuses on industrial ruination as a lived process. At any given moment, an industrial ruin appears as a snapshot of time and space within a longer process of ruination. Later, the ruin will inevitably undergo processes of demolition, reuse, or rebirth. Snapshots of industrial ruination can reveal a great deal about socio-economic processes, but cannot be separated from either the residential, commercial, community, and natural spaces in which they are located or the people who make up these surroundings. Thus, my analysis of industrial ruination as a lived process is framed within the broader context of landscapes and legacies.

Landscape and Place

The concept of "landscape" is a useful way to situate both processes and forms within "place" in the context of industrial ruination and urban decline. Landscape studies have encompassed a range of possible approaches including material, empirical, visual, and cultural approaches, and thus landscape provides a good framework in which to combine socio-economic and cultural analysis of industrial ruination. The cultural geographer Zukin (1991, 16) provides a useful definition of landscape:

> Landscape, as I use the term here, stretches the imagination. Not only does it denote the usual geographical meaning of "physical surroundings," but it also refers to an ensemble of material and social practices and their symbolic representation. In a narrow sense, landscape represents the architecture of social class, gender, and race relations imposed by powerful institutions. In a broader sense, however, it connotes the entire panorama that we see: both the landscape of the powerful – cathedrals, factories, and skyscrapers – and the subordinate, resistant, or expressive vernacular of the powerless – village chapels, shantytowns, and tenements.

Following Zukin, I define landscapes as an ensemble of material and social practices, and as symbolic representations of these practices. However, Zukin's stark division between the powerful and the powerless in her landscapes of "consumption and devastation" misses some of the complexity of landscapes, particularly in the context of lived experience. Moreover, the analysis does not apply to places of production because they always involve a mixture of labour, capital, and power and powerlessness.

There are some key tensions within the broad literature on landscape (which straddles art history, geography, archaeology, and cultural studies): tensions between distance and proximity, observation and inhabitation, and culture and nature. Wylie (2007) argues for an understanding of landscape that emphasizes lived experience rather than detached observation, derived from the work of the phenomenologist Merleau-Ponty (1989) and the cultural anthropologist Ingold (2000). In the phenomenological approach, "landscape is defined primarily in terms of embodied practices of dwelling – practices of being-in-the-world in which self and landscape are entwined and emergent" (Wylie 2007, 14). This approach contrasts with studies that neglect the human or "inhabited" dimension of landscape. Cresswell (2004, 10–11) is critical of the concept of landscape for this reason: he argues that "place" is a more inclusive concept, for place is inhabited, whereas landscape lacks people: "In most definitions of landscape the viewer is outside of it. This is the primary way in which it differs from place. Places are very much things to be inside of ... We do not live in landscapes – we look at them."

One of the aims of my research was to read landscapes of industrial ruination and urban decline as the products of social, economic, and cultural processes. The process of reading involved analysing snapshots of industrial ruination and physical evidence of urban decline across different moments in time, in order to discover what these material traces could reveal of wider social and economic processes. For example, the landscapes of industrial ruination betray certain clues to the time scale and degree of industrial decline, level of contamination, degree of state regulation, public uses, type of former industry, national context, and function within the wider social geography, among other indicators. Similarly, boarded-up houses, run-down shops, pocked roads, and other markers of urban decline are clues to the history of each place. I detail the methods, process, and implications of reading landscapes of ruination, deprivation, and decline in chapter 5.

While my research examines the spatial, material, and economic dimensions of landscapes of industrial ruination, it is also concerned with landscapes as inhabited places in which people live through processes of change. In addressing the latter theme, my research interrogates the relationship between people and places of ruination and decline over time. The concept of landscape does not adequately capture a sense of temporality, nor, according to Cresswell, does it necessarily encompass people. The related concept of "legacies" addresses this ambiguity and tension within the concept of landscape. Legacies of industrial ruination and urban decline refer to enduring features, both social and material, of former industrial and urban eras. The concept of legacies adds a stronger temporal dimension to the analysis.

Legacies

If landscapes of industrial ruination and urban decline are slippery and unfixed, then legacies are even more difficult to map. Legacies of ruination and decline are related to inheritance, historical traces, and generational change: the diffuse social, economic, cultural, psychological, and environmental impacts of industrial and urban decline on people and places. In all three case studies, I located people who were related to sites and processes of industrial ruination, either directly or indirectly, through residential or commercial proximity, present or former employment, interpersonal relationships, or involvement in community development. The most obvious and direct relationships between people and places of industrial ruination could be understood through examining the current uses of derelict sites. However, since most of the sites were abandoned or partially abandoned, relatively few people interacted with them physically. Some people worked in remaining industries, although a greater number had formerly worked in industrial sites in each case. The most-reported site uses were informal activities, such as vandalism, arson, drug and alcohol use, and theft of materials. One of the most important ways in which I explored people's relationships with sites was through their memories and perceptions, which – far from being merely subjective accounts – reflect divisions within social groups and between generations and classes. Perhaps most importantly, they also reflect much of the unease and difficulty experienced by people coping with transitions from an industrial past to an uncertain post-industrial future.

My analysis of memory and industrial ruination draws on the concept of collective or social memory. Collective memory, a term first coined by Maurice Halbwachs (1980), describes the shared and socially constructed memory of a group of people, as opposed to individual memory. The notion of collective memory has since been used in studies of national and public memories of traumatic histories, such as that of the Holocaust (cf. Williams 2007), as well as more generally in relation to complex processes of historical change (Blokland 2001; Nora 1989; Samuel 1994). Other scholars (Connerton 1989; Fentress and Wickham 1992) prefer the term "social memory" to "collective memory," which suggests a less homogenized, more complex interplay between the individual and the collective. Critics of the heritage and museum industries have also discussed the concept of collective memory. For example, Boyer (1994) argues that the contemporary postmodern city of collective memory is an artificial "museum" that consists of reinserted architectural fragments and traditions from the past. Some authors frame the split between official and unofficial memory in terms of an artificial divide between history and memory: Nora (1989) is critical of the split between "true memory" and historical studies of memory, and Samuel (1994) argues for a synthesis between history and memory and links the concept of memory to contemporary ethnography. My analysis of memory follows Nora and Samuel through conceptualizing memory as a dynamic and embodied force, but in the particular context of industrial ruination as a lived process.

Much of the literature on memory, and on industrial ruins in particular, relates to the social production of meaning and memory, both official and unofficial, and to its commodification in various forms of commemoration (Cowie and Heathcott 2003; Edensor 2005; Sargin 2004; Savage 2003; Shackel and Palus 2006). In an analysis of the relationship between social memory and industrial landscapes in the case of Virginius Island, part of Harpers Ferry National Historical Park in West Virginia, Shackel and Palus (2006) emphasize the struggle between labour and capital to control the meaning of the past. Another account of post-industrial conflict over meaning explores the contested politics of official memory-making and forgetting in Ankara, Turkey, since the 1950s (Sargin 2004). In an analysis of monuments to steel in Pittsburgh, Pennsylvania, Savage (2003) argues for the poetic and symbolic superiority of a slag pile (the waste product from the process of making steel) as a monument over officially sanctioned structures. Edensor (2005) also contrasts official memory with alternative memories. He highlights the recent trends towards "museumification" in the conversion of industrial ruins to homogenous sites of tourism and consumption, and goes on to explore "counter-memories" and "involuntary memories": multiple memories and forms of remembering stimulated by the "objects, spaces and traces" embodied in ruins.

I use the concept of "living memory," defined as people's present-day memories of a shared past, as opposed to official memory or collective memory. Living memory has diverse expressions across generations and social class, manifested through local experiences and practices in communities that are steeped in legacies of industrial ruination. Sites and processes of industrial ruination are deeply connected to the past and the memory contained within it, as they are physical reminders of industrial production and decline, and of the lives connected to them. The literature on nostalgia also suggests that memory is an experience of the present; nostalgia always tells us more about the present than it does about the past (Davis 1979; Shaw and Chase 1989).

Legacies include not only memories and perceptions but also numerous long-term socio-economic, cultural, and health impacts of industrial decline. In each case, I focus on the connections between the material landscape of decline and the adjacent neighbourhoods, communities, and cityscapes, and in particular on how these landscapes interrelate with social, political, and economic life. Landscapes and legacies of ruination and decline are deeply interconnected and cannot be separated. For this reason, I will address both landscapes and legacies together in the chapters that follow, focusing on the relationships between community, place, and social and economic change. At the heart of this analysis is a criticism of prevailing Western models of post-industrial development. "Post-industrial" refers to a socio-economic stage following the destruction of an industrial base, and is associated with decline in manufacturing in advanced capitalist countries and with the growth of knowledge, information, creative, and service economies (Bell 1973; Coyle 1998; Leadbetter 1998). However, this association is often more of an ideal than a socio-economic reality; many old industrial cities struggle to find new sources of employment and take divergent paths from the post-industrial mold. None of the case studies in this book are strictly post-industrial spaces; rather, they are in the process of moving from an industrial past towards an uncertain post-industrial future.

Methodology and Background to the Case Studies

The research for this book draws on three case studies of industrial ruination and urban decline, and employs the following methods: site and ethnographic observations; analysis of archival, documentary and photographic materials; "mobile methods" (see Büscher and Urry 2009) of spatial analysis, such as walking and driving tours of old industrial areas with research participants; secondary analysis of statistics; and, in each case, twenty to thirty in-depth qualitative interviews with a range of local people, including workers, residents, ex-workers, trade unionists, government officials, pensioners, urban planners, activists, and community and voluntary sector representatives. I conducted the research between 2005 and 2009, and devoted approximately two months of field research to each case study.[3] I have kept all interviewees confidential or given them pseudonyms to protect their identities, and conducted all research in accordance with the ethical guidelines of the Canadian Sociological Association, the British Sociological Association, and the Social Sciences and Humanities Research Council of Canada.

My methodological approach was inspired in part by Burawoy's (1998, 7) concept of the "extended case method," a reflexive and ethnographic approach to case study research which "thematizes our presence in the world we study" and explores connections between local contexts and global processes. According to Yin (1994), case studies are particularly useful for examining the complexities of contemporary real-life situations and processes, which corresponds to my interest in lived experiences of deindustrialization. I chose multiple case studies rather than single case studies to highlight the complex interplay between the local, regional, national, and global, and to draw out unique as well as cross-cutting themes.[4] However, the rationale for selection of multiple case studies in my research was not strictly comparative, at least not in the "comparative methodology" sense related to the scientific method of experimentation (see Ragin 1987). A more accurate way of describing my selection of case studies is "paradigmatic," a term Flyvbjerg (2001, 79) applies to describe a case study used "to develop a metaphor or establish a school for the domain which the case concerns."[5] I selected three paradigmatic case studies of industrial ruination and urban decline from three different regions of the world where deindustrialization has been most pronounced: the Rust Belt of North America, the North of England, and the old industrial regions of post-Soviet Russia. Each case study is in some ways typical of deindustrialization, yet each is also contextually specific or unique. The cases as a whole, through their combination of locally specific and typical features, are exemplary of cases of industrial ruination in an uneven geography of capitalist development.

My first criterion for selecting case studies was that each had to have experienced deindustrialization, but not full post-industrial transformation, and had to still have physical evidence of industrial ruins. This relates to my methodological choice to study places that have been "left behind" rather than the "success stories" of post-industrial transformation, such as Barcelona and Bilbao, which have been widely lauded as models of regeneration (González 2011). This choice also makes problematic the idea of "successful" post-industrial examples; the city centres of Newcastle upon Tyne and Niagara Falls, Ontario, are also commonly seen as success stories, yet a closer look at particular old industrial communities within both cities reveals evidence of industrial and urban decline. Each of the cities I selected was associated with a different iconic heavy industry: shipbuilding in Newcastle, textiles in Ivanovo, and chemicals in Niagara Falls. Each case was at a different phase of deindustrialization: Ivanovo was at an early phase; Niagara Falls was enduring a prolonged phase; and Newcastle was at a phase of impending regeneration. At the same time, the focus on old industrial sites and the communities surrounding them, rather than on the cities per se, grounded the research. The North American Rust Belt and Northern England were appropriate choices because both were from classic areas of industrial decline. The third case counterbalances the Anglo-American focus of (English language) academic literature on deindustrialization, and is situated in what one might call a rising star in the literature: post-Soviet Russia. The scale of industrial ruination in Russia is vast and much of its post-industrial geography undocumented, although this (documentation as well as geography) is rapidly changing. Cities such as Detroit and Manchester have become global symbols of deindustrialization, and of the limits and successes, respectively, of post-industrial transformation (Dicken 2002; Peck and Ward 2002; Persky and Wiewel 2000; Ward 2003; Zukin 1991). By contrast, my three cases have been relatively under-studied (but see Hang and Salvo 1981; Hudson 1998; Madanipour and Bevan 1999; Richardson et al. 2000; Shields 1991; Treivish 2004). Thus, this research contributes to specific knowledge about the social and economic geography and particularities of each case study in this context. By working comparatively, it also contributes to a wider theoretical discussion on how specific old industrial areas are connected to wider processes of social and economic change.

To summarize, the rationale for selection of case studies was based on the criteria that each case study would be: (1) an old industrial area that had experienced significant deindustrialization and had visible industrial ruins within its physical landscape; (2) on a different manufacturing-based industry, to show similarities and contrasts across industries, particularly different working cultures, gender dynamics, relationships with communities, and skill sets; (3) located in a different national context, to broaden the scope and potential for global comparisons of industrial capitalism; (4) at a different stage of deindustrialization, to reveal the complexities of deindustrialization as a temporal process; (5) located within a medium-sized conurbation, as these are more common than larger locations, yet are more heterogeneous than small mono-industrial towns; and (6) not as widely researched as other cities of industrial decline.

The research was theoretically driven, with the aim of generating ways to understand processes of industrial ruination throughout the globe. Together, the cases provide insights into how people cope with post-industrial transition in different national, political, social, and cultural contexts. With three case studies, it is possible to draw a range of comparisons: themes common to all three case studies; themes common to two case studies (with three possible combinations); and themes distinctive to each one (see table 1.1). If the study had used only two, distinctive insights would have been lost: the pernicious dimension of toxic contamination in Niagara Falls; the impacts of state regulation and protracted decline in Newcastle; and the post-Soviet context and the tenacity of old industries in Ivanovo. This research not only contributes to specific knowledge about the social and economic geography of each case, but also contributes to a wider theoretical discussion about common features, issues, and challenges in the landscapes and legacies of industrial ruination and urban decline.

I developed, adopted, and adapted the mixed qualitative case study methods of this research throughout the course of my investigation. On the one hand, I wanted to explore the theories of Harvey and Zukin on uneven geographies of capitalism, and to follow Smith and Harvey's (2008) invitation to examine uneven geographies of capitalist development at more varied and complex spatial scales. On the other hand, I wanted to remain open to what emerged from the data once I was in the field. As I moved through the different case studies, my work became increasingly ethnographic in focus. I asked research participants questions about the changes in the past twenty to thirty years in relation to jobs, housing, education, and social services; how they related to and identified with these processes and more specifically with the sites of industrial ruination; and about their individual and collective roles within change, contestation, redevelopment, or daily life. The ethnographic shift in my research occurred gradually as I started to incorporate material collected outside of the context of formal interviews, such as informal meetings, and driving and walking tours with informants. My analysis also shifted towards a focus on people's stories, and to the value of the narrative interview. The findings that emerged from the field data all came from the ethnographic observations and in-depth interviews, and included themes of conflicted place memory and nostalgia, attachment to homes and community, visions of community solidarity and contestation, uncertainty and stress, and legacies of industrial ruination that were particular to different forms of industry (toxic chemicals, shipbuilding, or textiles).

There were some limitations to the depth and scope of my case study research and analysis in these three sites. I was not able to spend an extended period in each location, so my information presents only a snapshot of time and space, and does not incorporate changes over a long period of time. However, this limitation fits with my analytical approach: to read slices of space and time in order to understand socio-economic processes at particular moments of industrial ruination. Each of the case studies had different challenges and specificities in terms of access, practical constraints, and richness of material, so they are not completely even, as in strictly comparative methodology (Ragin 1987), but rather illustrative and paradigmatic, with some scope for reflexivity, intuition, and adaptability within different contexts. Given the large amount of empirical data that I collected during my fieldwork, I had to be selective in which narratives I represented, particularly given the international and comparative scope of this study. I aimed to cover a wide range of perspectives and sources, yet I also had to be careful to maintain the richness of the qualitative material. In my data analysis, I reviewed all of the interviews and ethnographic field notes and identified key themes within and across case studies. Finally, I selected illustrative interview examples and ethnographic observations related to key themes within and across the case studies to discuss in greater depth. The narrative ethnographic approaches of Richard Sennett (1998) and Burawoy and Verdery (1999), which focus on particularly revealing or illustrative narratives from a selection of their research participants, inspired this methodological choice.

Case Study I: Niagara Falls, Canada/USA

The cities and highways along the North American Rust Belt were the original inspiration for this book. I selected Niagara Falls as a case study based on my memory of driving through the city as part of a cross-country road trip from Ontario to my home province of British Columbia in 2002. I was struck by the looming industrial ruins around Hamilton, Detroit, and Chicago, but I was even more astounded by the contrasts within Niagara Falls. In the public imagination, Niagara Falls is primarily a tourist destination, associated with honeymoons, casinos, and tawdry amusements. My parents spent their honeymoon there in the late 1970s, and I remember a childhood family visit in which I wore a plastic yellow raincoat and stood next to the falls. However, as I was to discover later on, beneath the myth of Niagara Falls on both sides of the border is a Rust Belt story of toxic contamination, boarded-up downtowns, and industrial decline.

Niagara Falls, Ontario, and Niagara Falls, New York, are twin cities built around the spectacular natural falls that straddle the international border between the United States and Canada. The Niagara region has almost two million regional inhabitants and seventeen million yearly visitors (Schneekloth and Shibley 2005). However, the twin cities themselves are relatively small: approximately 82,000 people live in Niagara Falls, Ontario (2006 Canada Census), and 51,000 live in Niagara Falls, New York (2008 US Census estimates), with opposing population dynamics of growth in the former and depopulation in the latter. Although Niagara Falls is best known as a tourist destination, the region has many other identities, as described in the following passage:

> Niagara certainly has its share of representations: the honeymoon capital of the world, one of the most lucrative gambling sites in North America, the great source of hydro power, the snowbelt of the US and the southern border in Canada, the rustbelt in the US and the wine country in Canada. Each of

Table 1.1. Key themes of case studies

Themes	Niagara Falls	Newcastle	Ivanovo
"Devastation but also home" (place attachment)	✓	✓	✓
"Imagining change, reinventing place"	✓	✓	✓
Spatialized socioeconomic deprivation and exclusion	✓	✓	✓
Politics of resistance	✓(Can)	✓	
Community solidarity	✓(US)	✓	
Strong collective "old industrial city" identities		✓	✓
Prolonged effort to sustain or re-build industry		✓	✓
Vast scale of ruination	✓		✓
Traumatic memory	✓		✓
Ambivalent nostalgia	✓		
Toxic legacies of contamination	✓		
Protracted, state-regulated industrial decline		✓	
Imminent regeneration		✓	
Tenacity of industrial and Soviet identities			✓
Pragmatism and functionalism in uses of old industrial sites			✓

> these imagined Niagaras sits in uneasy juxtaposition with the others; and each has consequences for the structure of governance, investment, and quality of life. (Schneekloth and Shibley 2005, 105–6)

The theme of "uneasy juxtaposition" is important when considering the social and economic landscape of the falls. Historical industries in the area include: tourism; steel; aircraft, mechanical, and electrochemical products; aluminium goods; and hydroelectricity, among others. Tourism – through casinos, honeymoons, cruise boats, and Disney-like amusements – has played a significant role in the historical and economic development of the region. Niagara Falls has been the focus of many studies in relation to its tourist industry, natural beauty, cultural significance, and position as an international border (Berton 1992; Irwin 1996; McGreevy and Merritt 1991; Shields 1991). The tourist industry has been more successful on the Canadian side of the border since the mid-twentieth century, partly because it has a better view of the falls, but also because of various political and economic factors. A physical landscape of abandoned and toxic industrial sites persists on both sides of the border, but is more prevalent in the US side. Tourist industries, abandoned and remaining heavy industries, dilapidated downtown centres, and vast stretches of natural beauty, among other contradictory features, comprise Niagara Falls. The case study of Niagara Falls, unlike the other two cases explored in this book, represents a "nested" or "embedded" case study of two cases within one case (Yin 1994) in order to capture the wider dynamics of industrial decline within the international border city-region.

The falls have long been exploited as a resource, both by the tourist industry, which has promoted them as a natural wonder of the world, and by heavy industry, which once thrived on the energy generated by the waterfall. The historic over-exploitation of Niagara Falls has been noted in the literature as a "tragedy of the commons," a phenomenon in which a public resource is subjected to overuse and underinvestment (Healy 2006; Ingram and Inman 1996). This theory describes the history of the tourist industry in Niagara Falls, which was unregulated in its early days in the late nineteenth and early twentieth centuries. There were warring hotel entrepreneurs, spectacles such as tight-rope walking, circus animals, a museum of curiosities from around the world, and "peddlers, hucksters, con artists and sideshow men on both sides of the falls" (Ingram and Inman 1996, 632). The tragedy of the commons is not invoked in this literature, as it could be, in reference to historical industrial over-exploitation.

The main reference point in the Rust Belt history of Niagara Falls is the infamous 1978 environmental disaster at Love Canal on the Niagara frontier in New York, when a toxic chemical dump was discovered underneath a residential neighbourhood and had disastrous health consequences for the community (Colten and Skinner 1996; Gibbs 1998; Mazur 1998; Newman 2003). However, Love Canal is just one of many examples of toxic contamination from the chemical industries in the Rust

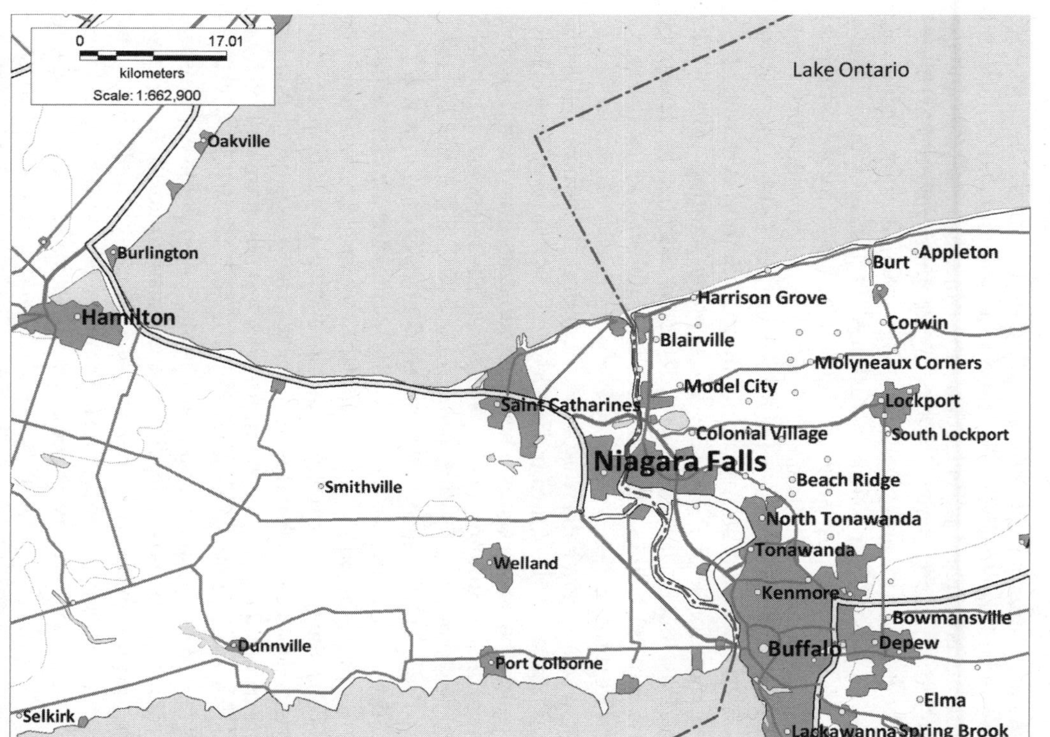

Map 1.1 Niagara Falls Region

Belt. There are numerous contaminated "brownfield" sites on both sides of the border in Niagara Falls. The twin cities are situated in different national contexts of deindustrialization – the American Rust Belt and the Canadian industrial heartland – which are sometimes referred to collectively as the North American Rust Belt (High 2003), as they form a vast region around the Great Lakes of rusted steel, car, and other heavy manufacturing plants. Canadian manufacturing activity, according to Lawrence McCann (cited in High 2003, 34–5), is concentrated in "a crescent between the western end of Lake Ontario from Oshawa to Niagara Falls [... and ...] a broad belt extending from Toronto to Windsor. These two zones comprise the Western Axis Manufacturing Area." The American Rust Belt is a familiar case of deindustrialization (cf. Bluestone and Harrison 1982; Cowie 1999; Cowie and Heathcott 2003), whereas the Canadian Rust Belt is less well-known.

My focus in Niagara Falls was on two abandoned chemical industrial areas – one on each side of the border – located in close proximity to low-income residential areas. I chose chemical industries from both countries to highlight cross-border parallels, particularly the relationship between contaminated sites and adjacent communities. Issues of social exclusion relating to race and ethnicity also emerged more strongly in this case than in others, for example in the case of the segregated African American community of Highland in Niagara Falls, New York. The case study of Niagara Falls offers distinctive insights into the legacies of toxic contamination as an aspect of deindustrialization. The inclusion of Niagara Falls, Ontario, in addition to the more widely known case of Niagara Falls, New York, as a single nested or embedded case study (Yin 1994) demonstrates the multiple manifestations and unevenness of industrial decline and its legacies within different national contexts. Ambivalent nostalgia and traumatic memory were key themes in the living memories of people in Niagara Falls, due to the impacts of job losses and toxic pollution, and the difficulty of separating positive and negative memories. The lingering health effects of pollution, combined with the continued socio-economic effects of unemployment and economic decline, produce a double burden associated with industrial ruination.

Case Study II: Newcastle upon Tyne, United Kingdom

It took me some time to select Newcastle upon Tyne as a United Kingdom case study, and I spent several months in preparation, researching different cities and taking train journeys to those I had shortlisted as potential cases, including Liverpool, Manchester, Stoke-on-Trent, Sheffield, Leeds, and Glasgow. I chose Newcastle partly because some of its old industrial areas remained untouched by regeneration, and partly because its deindustrialization had been studied less than that of other Northern industrial cities. I was also captivated by the atmosphere of the old industrial city built on the River Tyne: there was still a strong sense of local pride and collective memory based on shipbuilding. During my first visits to the city, I was struck by the juxtaposition between the regenerated quayside in the city centre and the abandoned shipyards to the east and west.

Newcastle upon Tyne is a city of approximately 220,000 people in the North East of England. Newcastle upon Tyne, Gateshead, North Tyneside, South Tyneside, and the City of Sunderland are the five local authorities that comprise Tyne and Wear, a conurbation of approximately one million people (see map 1.2). The 2007 English Indices of Deprivation reported that "the Region which has the greatest percentage of its LSOAs [Lower Layer Super Output Areas] that fall in England's most deprived 20% is the North East (34.2%)" (Noble et al. 2007, 12).[6]

The North East was built on coal mining, steel, engineering, and shipbuilding along the rivers Tyne, Wear, and Tees. The region, with Newcastle as its capital, reached its technological and industrial highpoint in the early twentieth century. It was one of the first regions in Britain to experience massive deindustrialization, yet this process has been prolonged. Newcastle's economy first suffered a major downturn in 1910 during a national slump that hit the coal and engineering trades particularly hard. Its economy revived during World War I, suffered again in the 1930s, revived briefly during the 1950s post-war boom, and has been in decline since the 1970s (Tomaney and Ward 2001). The abatement of manufacturing has been accompanied by limited service sector growth, and jobs in shipyards and factories have been replaced by jobs in call centres, night clubs, and shopping malls. The literature cites a number of factors that have contributed to decline in the North East, including strong foreign competition, overcapacity, labour conflict, government-led national economic restructuring (under Margaret Thatcher), the north-south divide in the United Kingdom, underinvestment and disinvestment, and poor business management (Charles and Benneworth 2001; Hudson 1998; Robinson 2002; Tomaney and Ward 2001). The majority of shipyard closures in Newcastle occurred between the 1960s and 1980s, but closures continued through the 1990s and into the 2000s. During that time, the national government

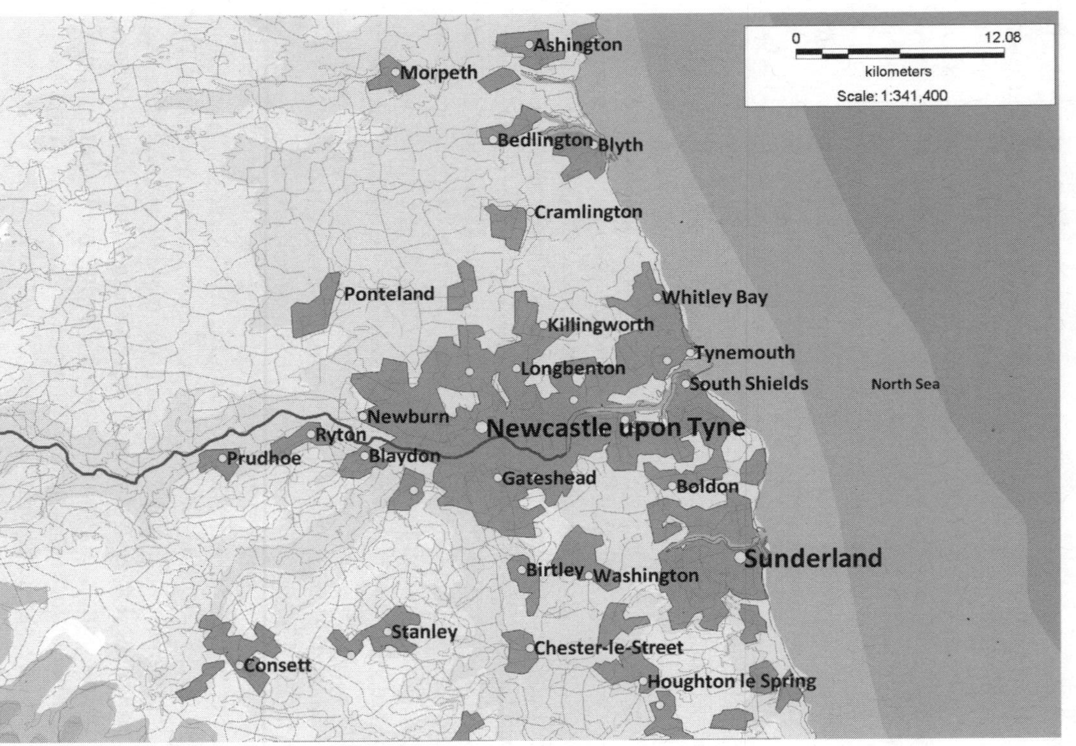

Map 1.2 Newcastle upon Tyne, North East England

offered lifelines to United Kingdom shipyards, such as Swan Hunter in Newcastle, in return for building government warships (Hilditch 1990), as well as for ship dismantling. Swan Hunter, the "last shipyard of the Tyne," closed in 2006.

In the 1990s, Newcastle attracted new industries, including call centres and inward investment branch plants (Tomaney et al. 1999; Tomaney and Ward 2001). This "new economy" has been the subject of much debate. Some critics have argued that the low-paid and female-dominated call centre jobs lack long-term prospects for economic growth (Richardson et al. 2000). Others have observed that the party culture in Newcastle and the Metro Centre, "one of Europe's largest indoor shopping and leisure centres" in neighbouring Gateshead, offer only limited regional economic growth, as they are based on consumption rather than production and much of that consumption is regional (Hollands and Chatterton 2002). The longevity of the new economy has also been questioned: the city experienced a wave of plant closures of new inward investment branch plants during the economic crisis of 1998, and many of the call centres that opened in the 1990s have since moved to India and other places in Asia.

In 1999, Newcastle City Council attempted to address some of the problems in the city related to deindustrialization, including significant depopulation, unemployment, and social and economic deprivation, by branding itself as "competitive Newcastle" and launching a controversial ten-year economic redevelopment plan entitled Going for Growth. This city-wide initiative sought to replace older housing with middle class homes in an effort to retain the middle-class population, which raised concerns that working class populations would be pushed out (Byrne 2000; Cameron 2003). With funding from the government's National Lottery and collaboration with the municipality on the opposite side of the River Tyne (Gateshead), Newcastle was remarkably successful at physically regenerating the city centre and the quayside area with riverside flats, restaurants, art galleries, and night clubs. However, other efforts were complete failures, such as the city council-led regeneration of Newcastle's West End, a deprived and disadvantaged former industrial area (based on engineering, armaments, and small-scale industries) that had been subject to repeated urban renewal policies since the 1960s (Madanipour and Bevan 1999; Robinson 2005).

When I first asked local residents about old factories in Newcastle, a typical answer was: "There are not many left. You should have come a long time ago." After many walks along the Tyne, metro journeys to

North Shields and the neighbouring rival city of Sunderland, and many queries with local residents, academics, and city council economic development officials, I decided to focus my research on the Walker Riverside industrial area and the adjacent community. Walker Riverside was one of the few remaining old industrial sites in Newcastle that was not regenerated at the time of my research, and it remained symbolically important as the former site of famous shipyards, such as the Neptune Yard and Swan Hunter. Deindustrialization in Walker has been long and protracted, processes of both decline and renewal have been mediated by the state, and the local community has resisted the city council's attempt at housing-led regeneration since 2001. This contrasts with the West End of Newcastle, which had already undergone several waves of demolitions and redevelopments over the previous forty years, and which no longer faced significant community resistance (Robinson 2005).[7] Imminent regeneration was a defining theme of the legacies of industrial ruination and urban decline in Walker. Many people no longer related to the shipyards directly, but instead felt connected to their homes and the community that had been built around, but was no longer itself related to, shipbuilding.

The case study of Walker, Newcastle upon Tyne, reveals social, economic, and spatial juxtapositions between sites of deprivation and regeneration within old industrial cities, and explores the uncertainties embedded in lived experiences of post-industrial change. Furthermore, it offers distinctive conceptual insights into the social impacts of protracted, government-regulated processes of change, as opposed to processes of unmitigated capitalist abandonment and redevelopment. To employ an analogy with the terms of post-socialist transition, this approach resembled "gradualism" rather than "shock therapy" (Dehejia 1997; Grindea 1997; Stephan 1999), which is ironic given the overarching context of Thatcherism in Britain. One of the criticisms of shock therapy with post-socialist transition was the lack of new institutions to replace those that had been dismantled overnight. In the case of post-industrial transformation in Walker, despite the long process, there was nothing to replace the industry that left. The proposed regeneration of Walker hinged entirely on property-led development rather than on local employment strategies. Even at the level of the city of Newcastle and the North East region as a whole, there was no substantive post-industrial economic engine to replace industry. This has strong parallels to the lack of post-industrial opportunities facing the post-Soviet textile city of Ivanovo, although the scale, scope, and speed of industrial ruination in Ivanovo were far greater, in part due to different starting conditions.

Case Study III: Ivanovo, Russia

Ivanovo is a post-Soviet textile city marked by extensive industrial ruination, urban decline, and deprivation. I selected Ivanovo as my third and final case study because the post-Soviet context offers a counterpoint to the Anglo-American focus of the other case studies, and to the broader English-language literature on Western deindustrialization. The study of post-Soviet landscapes of industrial decline provides a way to test the insights of political economy: how does a newly capitalist geography compare with older capitalist geographies? Does the analysis of an uneven geography of capitalism still hold? The textile industry had been heavily concentrated in Ivanovo, which was the primary textile manufacturer within the centrally planned economy of the Soviet Union. This concentration of industrial activity in one city-region as a starting condition before deindustrialization resulted in a much larger scale of industrial ruination than that in Western cases. The historic lack of economic diversification and Soviet logic of non-market-based enterprises also led many textile factories to re-open, without hope of profits, as a result of barter exchanges based on financial backing from suppliers and local political connections (see chapter 4).

In researching Ivanovo, I felt the most like a dereliction tourist: this was a source of unease when I spent my first days in the city wandering through a vast and unfamiliar urban landscape of industrial ruination. In fact, a description of Ivanovo in the "Way to Russia" online travel guide, written by a team of Russians to encourage tourism, describes "ruins" as the city's only tourist attraction:

> Ivanovo is a grey and gloomy city, with relics of the Soviet times on every step. It'll be enough to pass it through by bus going between Vladimir and Kostroma, just keep your eyes wide open: the central noisy and dirty street with grey residential buildings and a big red church in the middle of all the mess; the faded impressive mosaics to glory [sic] the Soviet heroes, left here from the 1970s; a dirty and noisy bus station with an old man playing accordion to cheer his fellow *babushkas*. (Way to Russia 2006)

The region of Ivanovo has a population of 1,148,329, and the city of Ivanovo, which is the region's administrative centre, has a population of 431,721 (Russian Federal Service of State Statistics 2002). Located approximately 300 kilometres northeast of Moscow, Ivanovo forms part of the "Golden Ring" railway network of ancient cities around the capital (see map 1.3). However, unlike the other cities in the Golden Ring – such as Yaroslavl, Kostroma, and Vladimir, which have fortresses, monasteries, and centuries-old white stone churches – Ivanovo is a modern industrial city with semi-abandoned textile factories and Soviet architecture. The only references to its pre-industrial history are dilapidated wooden country-style houses and old wooden churches. In 1871, Ivanovo-Voznesensk was formed as an administrative unit through the amalgamation of two villages, and it rapidly industrialized as a textile centre and earned fame as the "Russian Manchester." The population exploded from 17,000 in 1870 to 170,000 in 1917, making it the fastest growing city in Central Russia at the time. The city was populated by wealthy manufacturers and weavers in addition to a new working class, and the growing antagonism between labour and capital led to the first Soviet of Workers' Deputies in 1905 (Treivish 2004, 14). Ivanovo-Voznesensk reached the height of its textile production in the 1910s, but it only gained official city status in 1918, and was renamed Ivanovo in 1932. The revolutionary Soviet and industrial spirit of the city earned Ivanovo status as "the third Russian proletarian capital after Leningrad and Moscow," designated by Lenin himself. The city was also nicknamed the "City of Brides" or "the town of single women" (Browning 1992) because of its predominantly female workforce.

During the Soviet era, textile production in Ivanovo continued despite a steady decline in output, bolstered by the control of the centrally planned economy. In the 1960s, crane-building and machine-building plants were established in the city to improve the gender balance of labour and to diversify its mono-industrial structure, but this strategy was not very successful. Throughout the 1970s and 1980s, the textile workforce gradually declined. Ivanovo was one of the first cities in the former Soviet Union to experience deindustrialization. Within the literature, reasons cited for this include a dramatic decline in domestic demand and the low competitiveness of Russian industries, particularly light industries, in global markets (Kouznetsov 2004). The era of *perestroika*, or "restructuring" – a policy introduced by Gorbachev during the late 1980s that has been widely associated with the collapse of the Soviet Union – was a difficult period of social and economic transition in Ivanovo and in Russia as a whole. The social and economic impacts of the transition to market capitalism lasted for several years,

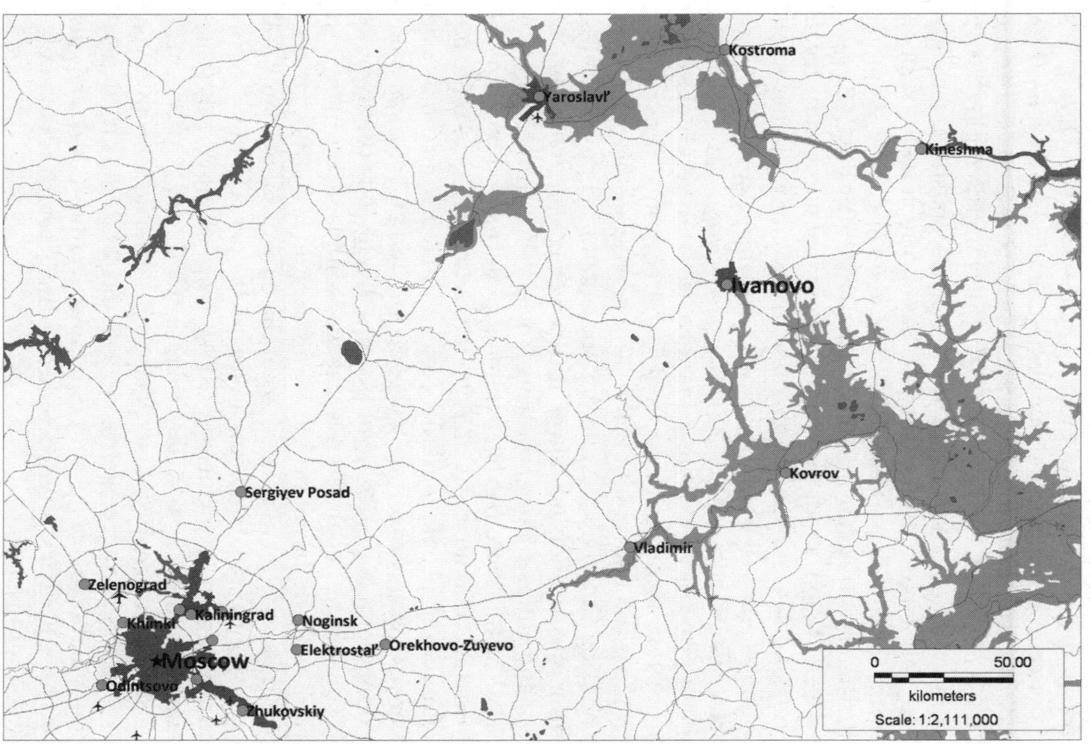

Map 1.3 Ivanovo and the Golden Ring of cities northeast of Moscow

with food shortages throughout the early and mid-1990s. During this time, 60 per cent of the Ivanovo population had to live on food from their gardens, a survival strategy in post-Soviet Russia known as the "dacha movement" (Burawoy et al. 2000; Sitar and Sverdlov 2004). Textile production in Ivanovo came to a complete standstill in the early 1990s. One of the largest textile plants, the Eighth March Textile Factory, closed in the 1990s and was converted into the Silver City shopping centre. Russian textiles made a significant recovery after the 1998 crisis, but there were doubts about the long-term survival of the industry in the early 2000s due to growing competition from China and other emerging economies and to the persistence of Soviet management practices within factories (Morrison 2008, 7–8).

Despite its proximity to Moscow, Ivanovo has fared poorly since the 1990s in relation to indicators of social and economic deprivation, even in comparison with cities in other Russian regions. The problems include housing shortages despite depopulation, high poverty rates, crime, drug and alcohol use, lower life expectancies, low quality of dwellings and infrastructure, and high levels of informal ("grey") business. Although there has been a nominal decrease in unemployment since the economic crisis of 1998, the figures are inaccurate because a number of people work for factories that are open only one month per year, wage arrears are still common, and the informal economy accounts for a large degree of peoples' livelihoods (Burawoy et al. 2000; Kouznetsov 2004). The average wage in Ivanovo is very low, and only a small number of people are affluent. In 2000, 10.9 per cent of people in Russia (on average) had an income of more than 4,000 roubles per month ($205), as compared with 0.1 per cent of people in the Ivanovo region (Kouznetsov 2004). The employment structure changed significantly with the collapse of industry (58 per cent of jobs were lost between 1980 and 1998), and available jobs are often given to men over women. Between 1990 and 2002, the city experienced depopulation of 6.8 per cent (Sitar and Sverdlov 2004), and this trend continues today as young people relocate in search of better employment opportunities.

One of the most extensive contemporary English accounts of the social and economic situation in Ivanovo is a Shrinking Cities Working Paper (Kouznetsov 2004; Sitar and Sverdlov 2004; Treivish 2004). Ivanovo was one of four urban centres that the Shrinking Cities Working Papers focused on, including Detroit, Ivanovo, Manchester/Liverpool, and Halle/Leipzig, in a project funded by Germany's

Federal Cultural Foundation. The project included exhibitions and publications about contemporary cities that have experienced "shrinkage" in population terms or decline in social and economic sectors, and had input from architects, academics, and artists. According to Treivish (2004), Ivanovo has three main stories: the paradoxical story of being located in the heartland of Russia, yet remaining marginalized; the Soviet and post-Soviet story of political ideals versus reality; and the typical deindustrialization story of long-term decline. Treivish describes a historical shift in Ivanovo from industrial growth to industrial ruins, from social revolution to social apathy, and from Soviet industrialist and constructivist urban planning to general decay and selective market-oriented renovation. Sitar and Sverdlov (2004) underline the specificity of the "Soviet socio-cultural model" as one of accelerated urbanization from an agricultural to an urban society which still maintains a level of continuity with the traditional principles of the "peasant world" and with the socialist past. My research is most concerned with Treivish's third story of "typical deindustrialization," although the other stories, particularly the specificity of the post-Soviet context, are deeply connected with this case.

Industrial ruination in Ivanovo was abundant and pervasive, as abandoned textile factories were scattered throughout the city, in addition to derelict and vacant houses and commercial buildings. At the same time, the landscape of ruination in Ivanovo was in a process of "reversal," as shown in the phenomenon of partially working and partially abandoned textile factories throughout the city. I decided to focus on the textile industries throughout the city as a whole, rather than those within a particular area, because the ruination of that single industry is widespread throughout the city. Pragmatism and functionalism emerged as themes that showed how people related to spaces of ruination, and how they lived and worked within a city characterized by significant industrial decline and limited urban renewal. Another theme that emerged was the tenacity of the textile and Soviet identities of the city. After the closure of virtually all textile factories in the city during the early 1990s, many factories gradually re-opened at a fraction of their original capacity. Despite the industry's ongoing struggle for viability, Ivanovo is still described within the official city literature as the "Russian Manchester." The case study of Ivanovo offers important insights into landscapes and legacies of industrial ruination and urban decline because of the significant scope and scale of its ruination and decline, because it is a recently deindustrialized city, and because of its post-Soviet context.

The Structure of This Book

This chapter has introduced the broad aims of this research project; the existing literature on landscape and place, legacies, deindustrialization, and industrial ruins; the theoretical framework of industrial ruination as a lived process; the mixed-method multiple-site case study research design; the background to the three case studies; and the methods of fieldwork and research analysis. The first part of this book – chapters 2, 3, and 4 – present each of the case studies of industrial ruination, community, and place. Rather than aiming for a comprehensive overview of a wide range of themes, these chapters highlight some of the most important and distinctive insights within each case while hinting at common themes. The second part of the book explores the conceptual themes that emerged in the research through comparative analysis of the different cases. Chapter 5 considers the theme of "reading landscapes of ruination, deprivation, and decline," and explores how one can read socio-economic processes within landscapes of industrial ruination and adjacent urban communities through a combination of spatial, visual, and social analysis. Chapter 6 explores the theme of "devastation, but also home" – many people who live in areas of industrial decline are attached to their homes and communities despite living among devastation. Chapter 7 expands the discussion to consider "imagining change, reinventing place" – how people cope with change and uncertainty, how they engage with the local politics of community and development, and how they imagine possible futures. The concluding chapter summarizes the central themes of the book, connects both distinctive and cross-cutting themes with theoretical debates, and reflects on some of the policy implications of this research.

Landscapes and legacies of industrial ruination and urban decline are evident throughout the industrialized world, and represent enduring and complex contemporary realities for people living through post-industrial change. Although my research is limited to three case studies, I could have selected many other examples. In fact, each person I tell about my research has a new story that relates to industrial decline; I have heard stories about the steel in Sheffield, the tin in Malaysia, the Ruhr area in Germany, the coal mines in British Columbia, and the vast industrial cities of China. It is my hope that the questions and themes explored in this book will resonate with myriad other places around the world.

PART ONE

Case Studies

Chapter Two

"When the Smell Goes, the Jobs Go": Ambivalent Nostalgia and Traumatic Memory in Niagara Falls

> We remember the bomb going off; we remember the Love Canal; we remember the plants closing; we remember our parents being out of work; that's what people remember. They remember all the fallout of the plants closing, the fallout from Love Canal. (interview with resident, Niagara Falls, New York, 21 March 2007)

Residents of Niagara Falls are still haunted by the traumatic 1978 experience of Love Canal, when a toxic chemical dump was discovered buried under the residential neighbourhood of LaSalle at the eastern edge of Niagara Falls, New York. Named after its creator, William Love, the Love Canal was an abandoned 3,000-metre canal in Niagara Falls that dated from the 1890s. In the 1940s, the Hooker Chemicals Company used the canal as a chemical dumping site for over 19,000 metric tons of toxic waste. After the canal was filled in 1953, the land was sold, and an elementary school and low-income residential houses were built on top of the site. The new residents were not aware of the area's toxic history, but they became suspicious when thick black liquid erupted in their basements and backyards. However, people only became seriously concerned when a number of illnesses, miscarriages, and deaths occurred within the community in the 1970s. Under the leadership of local resident and mother Lois Gibbs, residents campaigned to have the school closed and the health effects of Love Canal investigated. The residents faced an uphill battle against the opinions of scientific experts, Hooker Chemicals representatives, and government officials who wanted to contain the situation. The crisis escalated, with increasing deaths, illnesses, and media attention, which culminated when president Jimmy Carter declared of a state of emergency, closed the school, and evacuated residents in the homes on top of the dump.[1] The evacuated area still exists today as a large fenced-off empty field surrounded by blocks of abandoned homes, even though officially the Love Canal was declared safe in 2004 (Lane 2010). When I visited the area in April 2007, there were no signs on the fence to explain the enclosed barren field, nor was there any plaque or monument to commemorate the disaster (figure 2.1). This lack of public recognition seemed puzzling, given the scale of the disaster and the media coverage at the time. However, one needs only to look at the abandoned chemical factories lining Buffalo Avenue on the drive from downtown Niagara Falls to the northern suburb of LaSalle to realize that the scars of the chemical industry in Niagara Falls still have not healed. The past is not yet distant enough to enter the realm of official commemoration (cf. Boyer 1994).

Love Canal branded Niagara Falls as an international symbol of the human costs of toxic pollution. In the 1980s, Canadians for a Clean Environment, an activist group, guided busloads of people on "toxic tours" of contaminated industrial sites in Niagara Falls, Ontario, and Niagara Falls, New York. The tours were part of an effort to raise public awareness about the environmental contamination and health risks in the area. Everyone knew about Love Canal, but most people did not know that there were numerous similar sites that lay undiscovered. While the toxic tours took place in the aftermath of Love Canal, they were inspired by a toxic disaster known as the Chippawa Blob, which occurred on the Canadian side of the border in the early 1980s. The Norton Abrasives Company, the major employer in the town of Chippawa on the outskirts of Niagara Falls, Ontario, had regularly dumped toxic chemicals into the Welland River. The Chippawa Blob crisis began when people started to notice an oily film on the surface of the water:

> Over a three day period ... we had children suddenly showing up with burns, some kind of contact burns on their bodies, and their bathing suits were deteriorated, a couple of them had it on their faces and their hands, and then some of them were getting violently ill, very, very ill. (interview, founder of Canadians for a Clean Environment, 12 April 2007)

Canadians for a Clean Environment worked for six and half years with various agencies, government representatives, and citizens to clean up the Welland River. Their efforts cost over $3 million, and the Norton Company was fingerprinted through a link between dioxins and other compounds found both inside and outside of the plant. This case shows

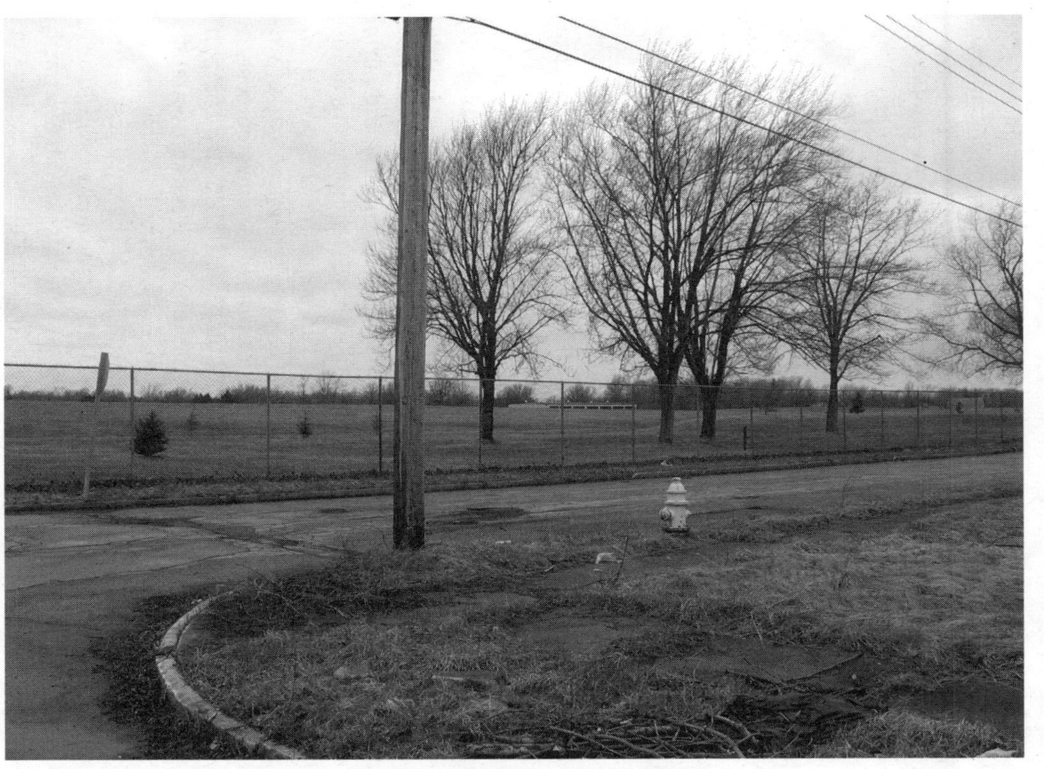

Figure 2.1 Love Canal site, Niagara Falls, New York, April 2007

that despite differences between Canada and the United States, the problems of toxic pollution and corporate negligence in Niagara Falls are by no means limited to the US side of the border.

Most of the chemical factories in the twin border cities of Niagara Falls have been abandoned, a process that began in the late 1970s and continues to the present day. The landscapes of industrial ruination in Niagara Falls, Ontario, and Niagara Falls, New York, were created by capital abandonment and followed widely documented patterns of deindustrialization across North America (Bluestone and Harrison 1982; Cowie and Heathcott 2003; Harvey 1999; High 2007; Persky and Wiewel 2000). However, capital flight in such a capital-intensive industry was not motivated primarily by a search for cheaper labour, but rather by falling profitability and markets, and tightening environmental regulations (cf. Colten and Skinner 1996; Hoffman 1999; Newman 2003). In fact, despite significant deindustrialization, a number of chemical companies remained in operation, albeit on a reduced scale, on both sides of the border at the time of my fieldwork in spring 2007. The decline of the chemical industries was related to their entanglement with the fortunes of other Rust Belt industries. Many chemical products were related to the war effort during World War II, and after it ended, the primary products of most chemical companies fed into the industrial process of steel and car manufacturing. As these neighbouring industries collapsed, and as companies moved in search of cheaper labour and resources, a considerable part of the regional chemical market was lost. Increasing environmental regulations and the associated costs of clean-up and investment in greener technologies also affected companies' profits. Love Canal and the Chippawa Blob raised public awareness of the heavy social and environmental costs associated with heavy industry, and that stigma damaged the reputation of the chemical industry as a whole. Despite some similarities, the geography of industrial abandonment differed between Canada and the United States, partly because of Canada's tighter environmental regulations and greater corporate liability, and partly because of the different relationships between industry and tourism on each side.

Some of the companies have faced legal action for negligent dumping, notably Hooker Chemicals in New York over Love Canal and Norton Abrasives in Ontario over the Chippawa Blob. However, most companies have not been held accountable for clean-up and remediation. A number of brownfields – "vacant or underutilized industrial and commercial properties where expansion or redevelopment is complicated by real or perceived environmental contamination" (Dunn and Dunn 1998) – are located throughout both Niagara Falls, New York, and Niagara Falls, Ontario. The remaining clean-up costs for contaminated sites involve so many variables that they are almost incalculably high, and have devolved onto governments and communities in both cities.

The social impacts of industrial abandonment for residents on both sides of the border include not only job losses but serious health problems associated with living and working near contaminated sites. In both cities, a disproportionately large number of people have reported illnesses such as cancer; heart, blood, and respiratory problems; and rare diseases. Residents of the nearby winery town of Niagara-on-the-Lake, Ontario (22.8 kilometres downstream from Niagara Falls), also reported unusually high numbers of miscarriages and illnesses during the 1970s and 1980s, which suggests that impacts of contamination spread to the wider Niagara region (interview, former Niagara-on-the-Lake resident, 26 March 2007). A great deal of general epidemiological literature emphasizes a correlation between living or working near to contaminated sites, and illnesses such as cancer (Brown 2002; Doll and McLean 1979; Draper 1994; Hang and Salvo 1981; Josephson 1983; Mitman et al. 2004; Tesh 1993). Hang and Salvo's (1981) influential book, *The Ravaged River*, provides epidemiological data that suggests a correlation between contamination and ill health in Niagara Falls, but it was written in the 1980s, and since then little research has gone into systematic epidemiological studies of the area. However, my research was not concerned with quantification of cancer clusters, as this is problematic even within the context of epidemiology (cf. Doll and McLean 1979; Draper 1994; Tesh 1993; Vrijheid 2000).[2] Rather, my concern was with qualitative interview-based material that illustrates the complexity, pain, and difficulty of lived experience with toxic legacies, and perceptions of the socio-economic and health effects of these legacies. Whether the perceived health effects in areas of industrial ruination derive from toxic chemical exposure, from the stress and fear of living in proximity to toxic sites, from other health problems prevalent within areas of socio-economic deprivation, or from a combination of perceived and real causes, remains uncertain.

I conducted research in Niagara Falls between March and April 2007, which included nineteen in-depth interviews with a range of people (local residents, former chemical factory workers, local environmental activists, local pensioners, church leaders, trade unionists, city council officials, and community development representatives), several group interviews with local residents and activists, local archival and documentary material analysis, driving tours with informants, and site and ethnographical observations. Residents' and workers' living memories – or present-day memories of a shared past (see chapter 1) – of industrial ruination, urban decline, and contamination in Niagara Falls were varied, and spanned indifference to sadness, nostalgia, regret, resignation, anger, denial, trauma, and acceptance. These differences could be traced across different generations, genders, classes, occupations, and ethnic groups. In Niagara Falls, Ontario, many local residents saw industrial decline as one story among many, rather than as a defining narrative for the city. Those who identified the toxic pollution from Cyanamid, the largest chemical factory in the city, as a significant narrative had direct connections to the factory, either through work or through residential proximity. By contrast, in Niagara Falls, New York, nearly all of the residents and former workers I talked to saw industrial and urban decline as the defining story of the city, though people who lived and worked in particular areas experienced the worst effects of job losses, poverty, and toxic contamination. Despite diverse living memories of industrial decline in both cities, the two key themes of ambivalent nostalgia and traumatic memory emerged in the accounts of those with the strongest connections to the landscapes and legacies of industrial ruination in Niagara Falls. This chapter will focus on these accounts, and will draw on illustrative interviews and ethnographic observations (cf. Burawoy and Verdery 1999; Sennett 1998).

Ambivalent nostalgia dominated many individual accounts due to the combined association of heavy industry with jobs and toxic pollution. The lack of a successor to replace the loss of industry factored into accounts of nostalgia among interviewees with the closest relationships to the industrial past. The bittersweet quality of nostalgia as a particular form of social memory is relevant within the context of ambivalent or contradictory memories. The philosopher Ralph Harper (1966, 120) points to inherent contradictions in the concept of nostalgia: "Nostalgia combines bitterness and sweetness, the lost and the found, the far and the near, the new and the familiar, absence and presence. The past which is over and gone, from which we have been or are being removed, by some magic becomes present again for a short while." Other authors (Davis 1979; Shaw and Chase 1989; Wilson 2005) have also argued that nostalgia is inherently paradoxical. The case of Niagara Falls takes this paradox one step further: in addition to the pain of longing for a different place or time, there is the pain of longing for a time of both positive and negative experiences.

Traumatic memory, epitomized by the experiences of Love Canal and the Chippawa Blob, was also a key theme for those people who were most affected by environmental disaster and industrial decline. Traumatic memory, or the memory of trauma (cf. Trigg 2009; Williams 2007), has been theorized primarily within the contexts of the Holocaust and wartime atrocities. According to Trigg (2009, 87), there is tension between place and trauma, as "sites of trauma articulate memory precisely through refusing a continuous temporal narrative." For Trigg, sites of trauma provide a link between subject, time, and place that is necessarily fragmentary, with tension between "a scene of recognition, in which specific details are recalled from the past and applied to the spatiality of the present" and a sense of displacement between recognition and experience, in which "the present is undercut by the radical singularity of the traumatic past, such that the simple fact of being there fails to contribute to the reality of time" (88). Disruption, contradiction, and tension were evident in the traumatic memories of people in Niagara Falls. These memories were revealed primarily in relation to sites and processes of industrial ruination: the fallout of Love Canal; the burns of the Chippawa Blob; the Cyanamid dust in Niagara Falls, Ontario, and the "black shiny stuff" from the Highland Union Carbide factory in Niagara Falls, New York, all stimulated traumatic memories, accounts of illnesses and deaths, and of the devastation of the industrial economy. Illustrative narratives of residents and former chemical workers in Niagara Falls will further explore the interplay between traumatic memory and ambivalent nostalgia.

In both cities of Niagara Falls, industrial decline and the lingering effects of contamination affected some people and neighbourhoods more than others, although the landscape of industrial ruination was more hidden in Ontario and more prominent in New York. In this chapter, I will focus on two of the most affected places within the toxic landscapes and legacies of industrial ruination, both located in close proximity to low-income residential areas: the former Cyanamid chemical plant in Glenview-Silvertown, Niagara Falls, Ontario, and the former chemical plants in Highland Avenue, Niagara Falls, New York. Addressing each place in turn, I will describe the wider urban context and explore people's memories, perceptions, and experiences of industrial ruination as a lived process.

Glenview-Silvertown, Niagara Falls, Ontario

Niagara Falls, Ontario, has a number of identities as a tourist destination, a port of entry, and an old industrial city. This Canadian city is less obviously associated with a stigmatized Rust Belt identity than Niagara Falls, New York, which remains branded with the memory of Love Canal. One exception within the literature is High (2003, 35), who identifies Niagara Falls, Ontario, as one of many areas within the North American Rust Belt that experienced plant shutdowns in the 1970s and 1980s. High also argues that despite widespread deindustrialization, Canadian workers rejected the negative image of the Rust Belt on grounds of nationalist economic ideology, and this "ideological rust-proofing ... mitigated the social and cultural effects of plant closings during the two wave of dislocation" (2003, 17). Indeed, the city's Rust Belt identity is buried beneath the gleaming surface of a tourist waterfront full of hotels, restaurants, casinos, and other amusements. Most visitors to Niagara Falls stay no more than one or two nights, never go beyond the Clifton Hill tourist area, and leave the city without much sense of its industrial past (interview, Niagara Falls City Council official, 22 March 2007). However, the abandoned power stations that stand just a few hundred metres beyond the falls offer some indication of the industrial history of the city. Before downtown revitalization efforts began, if a more inquisitive tourist drove into the historic downtown, its derelict state (figure 2.2) and closed main-street shops also signalled decline. Very few visitors venture further still, to the brownfield sites north of the historic downtown where Cyanamid once loomed.

The hidden old industrial identity of the city was evident in a conversation from the early stages of my research with Jack, the property manager of the Historic Niagara Development Corporation, which was situated in one of the many empty single-story offices in the dilapidated downtown. Jack came from Brooklyn, New York, and worked for a wealthy businessman in New York City who had purchased 60 per cent of the downtown properties with the aim of redeveloping the area. When I explained my research interests, Jack was incredulous. He had never heard of heavy industry on the Canadian side of the border, and he was certain that it was insignificant to the city as a whole: "It seems strange to me, this was never a heavy industry place like those places you mentioned ... the main industry in Niagara Falls is the tourist industry" (interview, 22 March 2007). Jack's perspective shows that to the outsider, the old industrial legacy of Niagara Falls is not well-known, even in the context of urban decline. At the same time, the New York developer's instincts for redevelopment potential were not entirely misplaced: the development of downtown Niagara Falls as an "arts and cultural district" has progressed since 2007,[3] following the model

Figure 2.2 Queen Street, downtown Niagara Falls, Ontario, March 2007

of arts- and culture-led regeneration discussed in chapter 1 (Bailey et al. 2004; Couch et al. 2003; Florida 2005; Landry and Bianchini 1995; Zukin 1982).

In many respects, the popular imaginary of the city is correct: Niagara Falls, Ontario, is a tourist destination with a diverse economy and is not a typical story of industrial decline. Between 1996 and 2006, its population increased from 76,917 to 82,184, which contrasts with the trend of depopulation in many old industrial cities, although the level of growth was lower than the provincial and national averages (2006 Canada Census). Heavy industries, including chemicals, silverware, plastics, abrasives, machinery, and paper products, have declined significantly in the past forty years, but the tourist industry has been relatively successful. A complex of hotels, various tawdry amusements, honeymoon activities, and other tourist commodities revolve around the falls, although the tacky and aggressive nature of the local tourist industry is not new but a continuation of historical tradition (Berton 1992; Dubinsky 1999; Healy 2006; Zavitz 2003). When the Ontario government introduced legal wagering in the local economy in the mid-1990s, there was a wave of further tourist income into the city. Its first casino opened in December 1996 and a larger casino followed in 2004, and the city currently faces gambling and addiction problems (Room et al. 1999; Turner et al. 1999).

Niagara Falls markets itself as a tourist destination and as a good place for international business. A review of the city council website (www.city.niagarafalls.on.ca) in 2008 found no references to decline in industries or manufacturing, and the environmental section of the website focused more on recycling policies than on contaminated sites or derelict industrial buildings. Rather, "vacant industrial sites" were promoted as "ready for building today at competitive prices" through public-private partnership deals and the promotion of industrial business parks. However, despite glossing over industrial decline in its marketing materials, the city has undertaken brownfield initiatives to address contaminated abandoned industrial sites. Brownfields can be found in particular areas of the city, some distance away from the main tourist area of Clifton Hill and from the green belt along the Niagara River.

The largest brownfield area identified by the city council is the Elgin Industrial Area, located next to (or within, depending on where the boundaries are drawn) the working-class residential community of Glenview-Silvertown, and near to the historic downtown. The Elgin Industrial Area has been the focus of brownfield redevelopment plans since 1993 and was selected as a pilot project area by the city in 2005, based on "the number and large size of properties in this area, proximity to the Downtown, and significant land use changes occurring in the area" (RCI Consulting 2006).[4] I focused my attention on the former Cyanamid plant in the area, as Cyanamid was the largest chemical factory in the city's history. The research draws on accounts from people who had worked in Cyanamid or lived in Glenview-Silvertown within close proximity of the plant.

The Glenview-Silvertown neighbourhood in Niagara Falls has no fixed name, and different sources refer to it as "Glenview," "Silvertown," "Elgin," or even as part of the old downtown. In the past, people referred to the area as Silvertown due to its early industrial history, but more recently, some people have preferred to use Glenview because Silvertown is negatively associated with industrial decline (RCI Consulting 2006). Many people use the hyphenated Glenview-Silvertown to capture both new and old (or popular and official) identities of the area. The Niagara Region municipal government does not identify Glenview-Silvertown as a neighbourhood in its own right, but rather places both it and the historic downtown within the Elgin neighbourhood, for a total population of 11,585 (Niagara Region 2006). Elgin is one of the poorest areas in the city, and has six out of seven indicators of socio-economic deprivation that rank well above (i.e., worse than) the municipal average (Statistics Canada 2006a). For example, 16 per cent of people in the neighbourhood are below the "low income cut off" – a Canadian socio-economic indicator of low income based on family expenditure data where families devote a much larger share of their income on basic necessities of food, shelter, and clothing than the average family – as compared with a municipal average of 9.6 per cent. Furthermore, the neighbourhood has higher percentages of children born at risk of development disorders, people who have not completed high school education, vulnerable children, unemployment, and families who spend more than 30 per cent of their income on rent. Demographic data for the 2006 Canada Census Tract 0215.00, which most closely corresponds to the neighbourhood of Glenview-Silvertown, shows depopulation of 1.3 per cent, from 4,449 in 2006 to 4,392 in 2006, and higher unemployment, at 7.5 per cent as compared to the municipal rate of 6.2 per cent (Statistics Canada 2006b). The relative socio-economic deprivation of the Elgin (and Glenview-Silvertown) neighbourhood suggests that there is a significant uneven geography of development, not only between cities, but also within them (Massey 1984).

Cyanamid, an American company that produced calcium cyanide and other chemical products, was once the largest manufacturer and biggest single employer in Niagara Falls, Ontario. Its main plant was located on Fourth Avenue in Glenview-Silvertown. The plant was founded in 1907 as a fertilizer producer; shifted to war industry production of ammonia, ammonium nitrate, and special chemicals during the 1940s; and resumed the manufacture of fertilizer and cyanide-related products after the war. Cyanamid also provided a range of services and benefits for its employees, and fostered a sense of solidarity and community. Such activities were typical of the benefits many companies offered their employees during the Fordist era (1930s to 1960s in North America), which was based on mass production, Taylorist managerial practices, and mass consumption (Amin 1994; Boyer and Durand 1997; Harvey 1999; Jessop 1991). For example, Cyanamid opened an employee recreation centre in 1951, and later donated it to the YMCA. The company also built a large and popular swimming pool on its grounds that remained open from 1936 until 1971. The pool closed when it became unprofitable, largely because it did not meet new government regulations that stated that one had to be able to see the bottom of the pool (Ricciuto 1995).

Although Cyanamid had a generally positive local reputation during its many years of operation, it faced problems with industrial action and environmental regulations. Periodic labour disputes and strikes over wages occurred between the 1950s and the 1980s. On several occasions, Environment Canada updated regulations concerning furnace emissions and waste disposal, and Cyanamid was forced to comply by upgrading its equipment. Glenview-Silvertown residents complained that "Cyanamid dust" covered their cars, houses, and laundry. When the plant closed in 1992, newspaper reports reflected a sense of loss on the part of the community over the factory's decline (Murray 1992; Skeffington 1992). A follow-up newspaper report was done a year later to document the unemployment the closure had caused (Skeffington 1993). The factory was finally demolished in 1995 after Cyanamid failed to sell it, and the vacant field in which it once stood is now an unseen ruin where only traces of contamination and memories remain (figure 2.3). The only building left when I visited the site in 2007 was the boarded-up YMCA, which was pivotal in local struggles over the legacies of toxic contamination (figure 2.4).

The other Cyanamid plant, located in the south end of the city, remained in operation during the time of my research, although at a significantly reduced capacity of a few hundred people and under a new company

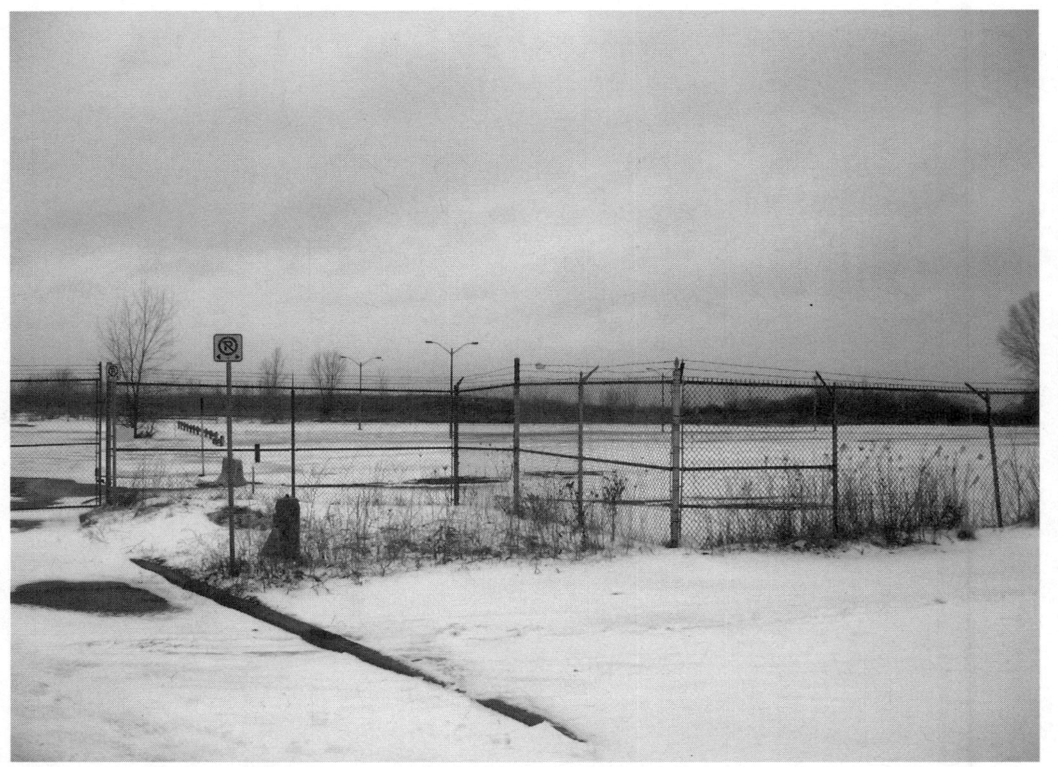

Figure 2.3 Former Cyanamid site, Niagara Falls, March 2007

Figure 2.4 Former Cyanamid community centre and YMCA, Niagara Falls, March 2007

name: Cytec Industries. Cyanamid changed its name in 1991 to distance itself from the legacy of environmental contamination, as its original name bears a close resemblance to the word "cyanide." Cytec is a global specialty chemicals and materials company that specializes in end markets in aerospace, adhesive, automotive and industrial coatings, chemical intermediates, inks, mining, and plastics. An analysis of the company website in March 2011 showed that Cytec has worked to build an image of corporate social and environmental responsibility, and places "safety, health and environment" at the top of its list of company values (Cytec Industries 2009). Following the closure of the Fourth Avenue Cyanamid plant in 1995, Environment Canada required that Cytec pay for some of the clean-up costs. The company has since studied the contaminants in the soil and performed limited remediation efforts. However, concerned environmentalists and residents have argued that these efforts are insufficient, particularly since the company has not permitted independent testing of the site. Debates as to whether the site has enduring health implications for nearby residents in Glenview-Silvertown persist today.

In 2006, Niagara Falls put forward a brownfield redevelopment proposal to build a community skating arena in the place of the old YMCA on the former Cyanamid site. Many local residents felt that a new arena was badly needed, for there were no community or recreational centres in the downtown area – nothing had replaced the YMCA when it closed. However, some residents, who were concerned about the health effects of residual contamination and claimed that the site had not been properly tested or cleaned up, were strongly opposed to the proposal. Their case received backing from Unite Here, a national trade union that represents hotel, casino, and other service sector workers. A number of the trade union's members live in the Glenview-Silvertown area because of lower housing costs, and Unite Here was concerned about the health and safety of its members, not only in the workplace but in their homes (interview, Community Organizer for Unite Here, 5 April 2007). After a series of debates at the Niagara Falls City Hall in 2006 and 2007 that involved residents, corporate representatives, lawyers, trade union representatives, city councillors, and scientific experts, the Niagara Falls city council voted to proceed with the arena plans, and construction began in October 2008.

The interplay between past, present, and future in the former Cyanamid site reveals many of the complexities of processes of industrial abandonment, ruination, and redevelopment. The physical ruins of the factory are gone, but evidence of ruination remains: a vast open field surrounded by fences, unknown levels of contamination, and a boarded-up community centre (later to become a building site in 2008, and later still a skating arena). This case also shows the difficulty of measuring and assessing the legacies of toxic contamination. The founder of Canadians for a Clean Environment, one of the leading environmental activists in Niagara Falls, observed that Cyanamid was only one of many polluting local industrial companies, and most of the damage it did went into the ground and the river (interview, 12 April 2007). He argued that despite the recent controversy, the site does not currently constitute a health risk and there are more pressing local environmental issues. But even if one were to accept that the site is no longer a threat to human health, what about the health problems of the people who grew up around the area, or those of the people who worked in the factories? What about the long-term impacts for the community that lived alongside toxic pollution for several decades? What about the symbolism of the place: what does it mean for public memory?

Remember the Cyanamid Dust?

My analysis of lived experiences and memories of industrial ruination and urban decline in Niagara Falls, Ontario, draws on different narrative accounts that provide illustrative contrasts among those who remember the "Cyanamid dust." Most local residents were aware that Niagara Falls has a history connected to heavy industry. However, the relative importance of industrial history in relation to other city histories differed according to age, social and family background, area of residence, and employment history. Former factory workers, workers' families, and residents who grew up near the factories identified most strongly – whether positively or negatively – with the city's industrial past. However, memories and experiences of industrial decline and pollution were mixed: some interviewees had wholly negative reactions because of the health effects associated with pollution, while others also remembered economic stability, worker solidarity, and community spirit. This division suggests that class, as related to employment and residential community, affects identification with industrial history (Burawoy et al. 2000; Cowie and Heathcott 2003; Milkman 1997; Roberts 1992).

At the time of my research in spring 2007, Jim – a community activist and former Glenview-Silvertown resident and steelworker – had been campaigning for the clean-up of the former Cyanamid property for over sixteen years. He had suffered from respiratory problems as a child, had three generations of cancer within his family, and had noticed that there was a local concentration of serious illnesses. In 1999, Jim addressed an open letter to the people of Glenview-Silvertown, entitled "Remember the Cyanamid dust," in which he asked residents to come forward with stories of health problems that they felt might be related to living near the plant. By collecting these stories, Jim wanted to prove there was a local cancer cluster that was a direct result of Cyanamid's industrial operations. The letters and emails he received detailed illnesses such as cancer, respiratory illnesses, and blood diseases. Many responses also included personal accounts of the Cyanamid dust, which, for Jim and several other residents, was the main memory connected with the plant:

> I remembered the Cyanamid dust because I used to wash my father's car in vinegar to get [it] off ... In many instances, it left spots that wouldn't wash out. My mother would have to choose when to hang the clothes on the line because of which way the wind was blowing. When the wind was blowing in the wrong direction, you didn't do your laundry that day. (interview, Jim, 27 March 2007)

Jim presented his campaign to Environment Canada, along with the documents, reports, photographs, and letters he had collected. In response, Environment Canada conducted a study of the local health impacts of the former plant, which relied on site tests preformed by Cytec consultants and concluded that no health effects were associated with living near the site.[5] According to an article in the *Niagara Falls Review* (Pelligrini 2005), a local newspaper, Environment Canada's report argued that several alternative factors could account for the health problems in Glenview-Silvertown: "Lower socio-economic status, higher levels of smoking or alcohol consumption or the old-fashioned way in which houses were heated – by burning oil or coal – could be the culprit." The report also noted that other previous industries in the area, such as the silver manufacturers of the 1800s, could have caused the problems (Pelligrini 2005). The possibility of a correlation between health risks and heavy industries was thus implicitly accepted, but only in the context of industries that had existed prior to Cyanamid. Jim encountered many obstacles over the course of his campaign, but at the time of our interview he planned to continue it. He expressed nothing but regret and anger over the toxic legacies of Cyanamid, and his account reflects traumatic memory rather than ambivalent nostalgia.

Other people remembered Cyanamid, despite its pollution, with sadness for the loss of industry. One newspaper article described receiving "nothing but fond recollections" when local residents remembered the Cyanamid pool (Ricciuto 1995). Another local article reported: "While the citizens complained about the dust from the Fourth Avenue plant, or ammonia from the Welland site, thousands of families were fed, housed, clothed and educated on Cyanamid paycheques" (Murray 1992). In "The Cyanamid Closing: A Year Later" (Skeffington 1993), there was a follow-up report on the 240 employees who lost their jobs when the Cyanamid plant closed in 1992. According to statistics from a job placement centre set up by Cyanamid, 123 people had found full-time employment. Approximately 90 workers remained unemployed, and the article followed up on their accounts of struggle in the labour market. I spoke with one of the workers who had been interviewed, Ben.

Ben was forty-seven when Cyanamid closed, and he had difficulty finding new work because of his age. It took him six years to obtain full-time employment, and he worked odd jobs during that time to get by. Ben's loss of secure employment put stress on his relationship with his wife, and they argued about money. He suffered from serious lung and breathing problems related to his factory work; he said that the inside of his nose was "burnt [and] scarred" and that he had to use a respirator to breathe at night. Both of Ben's sons had also suffered from health problems; one had cancer and another required a liver transplant. Ben acknowledged that there might be a connection between his second son's cancer and the environment:

> The area of Niagara Falls, Niagara region, it's a very high cancer occurrence over here. My second son, he was eighteen, he had testicular cancer. He was lucky that he caught it when he did because the doctor told me that if it had been another month then he wouldn't have survived. But that's his fault because he didn't sight it. You know, when you're young you don't want to know what's happening, you ignore it and it will go away. He was only eighteen. So what caused it, I don't know. I'd guess, quite often, it was the environment.

Although Ben felt that the health and environmental working conditions at Cyanamid were bad, he was reluctant to blame the company for his sons' health problems, and instead framed them as caused by the environment in general. Ben stressed that Cyanamid was a decent employer. He regretted the company's closure, and noted that it had been important to the city both economically and socially:

Q: How did you feel when the plant was demolished?

BEN: It was sad. Cyanamid was a good thing for the community. One of the policies they had, they hired employees' children, school age kids, and there were a lot of kids in Niagara Falls who got their university education from Cyanamid. The overtime in Cyanamid was phenomenal. I worked there for ten years, and in the first eight years I had one flat cheque [cheque without overtime].

Q: Did you like to work overtime?

BEN: Overtime is part of the piece.

Q: Did you enjoy the work?

BEN: I enjoyed the work. I enjoyed the fellowship for the employees. As far as the environment went, no I didn't enjoy the environment. There was no heat in the plants, so it was sometimes too hot, sometimes too cold.

Ben's memories of the factory contain a mixture of positive and negative elements. He remembered the camaraderie between workers, the sense of community, and the high wages with fondness. Yet he expressed sadness at the loss of industry and at the health effects that both he and his sons have suffered. Although both Jim and Ben were residents and factory workers, and both reported health problems that they connected with Cyanamid, Jim felt deeply wronged, whereas Ben accepted the environmental risks as part of heavy industrial work. Ben's account reflects both ambivalent nostalgia and traumatic memory over the industrial ruination and toxic legacies of Cyanamid.

Sharon, another downtown Niagara Falls resident, situated the industrial past in greater tension with the tourist industry of the past and present. Sharon was in her sixties at the time of our interview and had worked in business redevelopment in downtown Niagara Falls in the 1980s. She noted that both she and her husband had a family background in tourism rather than in industry. She stressed that although her husband was a school principal, he had worked hard to obtain his teaching qualifications, and his family, who came from Hungary, had also worked hard as migrants to the area. In this way, Sharon set her family apart from people she perceived as being less positive about change and less hard-working. For example, she mentioned the lack of business acumen of local retailers as a factor in the decline of the downtown area. She argued that most people in Niagara Falls were unjustifiably negative about the tourist industry, which created good jobs, opportunities for community redevelopment, and economic vitality. She compared the tourist industry to Cyanamid, and noted that, like Cyanamid, tourism had become the main source of jobs and tax revenue in the city. When I asked whether she thought people in the local community felt sad when the factory closed, she replied: "Those that worked there. The rest of us thought ... all this pollution. No, only those that worked there, which is natural" (interview, 27 March 2007). Sharon, who had not felt sad herself, was not connected to industry by either employment or family background. The industries fit into her account as part of the history of Niagara Falls, but not as part of its present or future. Although she was not nostalgic about Cyanamid's closure, she revealed frustration, loss, and anger at the decline of the downtown, for which she blamed individual choices and work ethic rather than structural economic factors. The tensions in her account between lack of regret about industrial decline on one hand, and regret about the decline of the downtown on the other, suggest a different form of ambivalent nostalgia in her personal experience of post-industrial change.

In Niagara Falls, Ontario, local residents' memories and experiences of landscapes of industrial ruination, exemplified by the Cyanamid case, reflected the themes of ambivalent nostalgia (Davis 1979; Harper 1966; Shaw and Chase 1989; Wilson 2005) and traumatic memory (cf. Trigg 2009; Williams 2007) in different ways. Jim's account reflected health-related trauma rather than nostalgia, while Ben's account reflected both trauma and ambivalent nostalgia, as his memories were more positive despite the health consequences for him and his family. Sharon's account reflected ambivalent nostalgia but not trauma, and she framed her sense of loss and frustration in relation to the decline of downtown, which she saw as unrelated to the decline of industry. Following Trigg's (2009) analysis of trauma and memory, the traumatic memory accounts "refused a continuous temporal narrative," with contradictory, disrupted, and fragmentary elements. Jim's account was in some senses frozen in time, a constant, polished re-telling of his own story and others that has lost its narrative dimension through repetition. Ben's narrative interspersed fragments of trauma, such as his sons' ill health, with positive memories. Each narrative reflects the tensions of lived processes of industrial decline: tensions between health and justice; between health and economic prosperity; between industry and tourism; and between different socio-economic classes. While the landscapes and legacies of industrial ruination – evident in the brownfield sites in the historic downtown and the living memories of residents and workers – remain submerged beneath the tourist identity of Niagara Falls, Ontario, the rustbelt identity of Niagara Falls, New York, to which we will now turn, is plainly visible.

Highland Avenue, Niagara Falls, New York

Over the last forty years, the population of Niagara Falls, New York, has declined steadily from over 120,000 people in the 1950s to 55,593 in 2000 and to 52,326 in 2006 (US Census Bureau 2006). More than 200 companies, primarily based in heavy manufacturing industries, left the Niagara region between the 1960s and the 1990s (Newman 2003, 128). Historical industries in the area include steel production, aircraft, mechanical and electrochemical products, aluminium goods, and hydroelectricity. The tourist infrastructure is neither as successful nor as extensive as it is in Ontario. The most recent tourist development at the time of my research was the Seneca Niagara Casino and Hotel, which opened in 2002 and was the result of the US$80 million transformation of the former Niagara Falls Convention Centre. The casino is one of the city's largest employers, but it faces racial tension because many older residents resent the fact that it is owned and operated by the Seneca Indian Reservation (group interview, Seniors' Centre, 12 April 2007). Half of the infrastructure in the city, including not only factories but also roads, housing, sidewalks, and commercial buildings, is in a state of disrepair. When I first crossed the border from Niagara Falls, Ontario, into Niagara Falls, New York, the border guard asked me what I was doing, and when I answered "researching industrial decline," he said that I had come to the right place because unlike in Canada, the United States "did it the wrong way" and did not keep tourism and industry separate.

There is significant environmental devastation in Niagara Falls, New York. Billions of gallons of chemicals have been poured into the river. In the 1960s, the Chemtol Chemical Corporation dumped hazardous chemical waste into open-air lagoons near the Niagara River shoreline, and in the 1970s, Stauffer Chemical Corporation dumped toxic waste directly into the river. Chemical dumping continued through the 1980s, and it has been recently estimated that 90 per cent of all toxic waste pollution in the area comes from chemical dumps that were abandoned by former industrial and chemical companies, including paper companies, chemical corporations, generating stations, metal companies, and chemical waste companies (Berketa 2005). The United States Environmental Protection Authority within Niagara County has identified more than twenty large-scale brownfield hazardous waste sites, seventeen of which are in the city of Niagara Falls itself.

As of spring 2007, the Niagara Falls, New York, website acknowledged the city's economic difficulties, noting that it had lost two-thirds of its industrial workforce in the past forty years (Niagara Falls Empire Zone 2007). Since this time, the city council has developed a new website which, like that of Niagara Falls, Ontario, glosses over such problems and focuses on opportunities for investors and visitors (City of Niagara Falls 2011). At the time of my research, the city offered business incentives for economic development, and had a brownfield redevelopment initiative managed through the Office of Environmental Services. The initiative aimed to test sites for contamination and work towards remediation of selected brownfield sites. According to the senior planner for the City of Niagara Falls (interview, 21 March 2007), the funds for brownfield redevelopment were limited: enough for initial site investigations to assess levels of contamination, but not enough for substantial reme-diation. In 2007, the city council's website noted that joint efforts to address brownfield sites between the Canadian and United States governments have been underway since the 1980s. However, according to interviews with city council officials on both sides of the border, there had been little cross-border communication or collaboration on these issues as of the time of my research.

There are two main brownfield areas in Niagara Falls, New York. The largest is the sprawling Buffalo Avenue industrial corridor, which runs east along the Niagara River for several miles, and although many of the factories are abandoned or demolished, several still operate at a reduced capacity. The senior planner for the city council explained that many of the factories along Buffalo Avenue are nominally in operation because companies prefer to keep their properties and pay taxes, rather than sell them and risk legal responsibility for clean-up (interview, 21 March 2007). The second main brownfield area in Niagara Falls is located further north up the river, in the Highland Avenue community. I decided to focus on this area because of the dynamics of spatialized racial and social exclusion, the lack of existing studies on the community, and the relatively small size of the industrial corridor as compared with Buffalo Avenue.

Highland Avenue is one of the poorest communities in Niagara Falls, New York, and is a largely African American area on the north edge of the city that is physically set apart by railway tracks on one side and the city border on the other. The area developed after World War II, when jobs in chemical factories were booming and African American people from the South were moving to find work. They were not allowed to live downtown due to racist city policies at the time (see chapter 5 for a discussion of "environmental racism" and spatialized racial and social exclusion) and were instead pushed into housing projects directly next to the chemical factories in Highland. They moved to this area and gradually built their own homes and community. As industry declined, many people left, and many of those who remain are widows whose husbands died prematurely from working in the plants. Highland Avenue has a population of approximately 3,000, and its eighteen churches are the focal point of community networks and support, based on African American Baptist religious traditions brought from the South.

According to a study of access to employment for adults in poverty in Niagara Falls and Buffalo based on 2000 US Census data (Hess 2005), Highland Avenue is the poorest neighbourhood in Niagara Falls; with 64 per cent of adults living in poverty, a median annual household income of US$8,470, and the highest percentage (89 per cent) of African American residents in the city. The next two most impoverished neighbourhoods in Niagara Falls are the downtown and the mid-city; both neighbourhoods have 44 per cent of adults living in poverty and median annual incomes of less than $16,000, and have African American populations of 51 per cent and 62 per cent respectively. The study reveals the wider regional geography of severe racial segregation in both Niagara Falls and Buffalo, and shows how poverty in each city is concentrated relatively near to the downtown core.

There are four abandoned chemical plants in Highland Avenue that are in close proximity to residences, and a few local factories are still in operation. The Highland Avenue brownfields exhibit greater neglect and more uncertain contamination levels than the Cyanamid site. One of the largest is the Highland Avenue Union Carbide site, which is situated on eighty-eight acres of overgrown land. Union Carbide was one of several large chemical companies that operated in Niagara Falls, and the national plant on Highland Avenue was one of at least six company plants that operated in the region during the heyday of the chemical industry in the 1950s and 1960s. Union Carbide was an American multinational company, incorporated in 1917, that produced over 700 different chemicals. The Niagara Falls Union Carbide plants specialized in chemical production during World War II (including work on the Manhattan Project) and for the steel and metal industries. Labour conflicts within the plants revolved around issues of race as well as class. In the 1970s, there were several reports of racism within the Union Carbide plants, and the Black Employees Club in the carbon products division in Niagara Falls accused the company of maintaining a "prejudiced, racist and biased system" and said that "minority employees are limited and restricted as far as promotion, job-upgrading and training as compared to white employees" (Summers 1971). Union Carbide reduced its operations in Niagara Falls in the 1980s, and had only two plants, including that in Highland Avenue, by 1985.

Niagara Falls City Council adopted the Highland area as a pilot project when the vast Union Carbide factory was abandoned in the mid-1990s.[6] According to the city council's senior planner (interview, 21 March 2007), the city still intends to investigate, clean up, and redevelop the site, but lacks the funds and the economic interest from developers to do it. The dynamics of the Union Carbide site are similar to the Cyanamid site, in that the respective city councils have acquired ownership of all or part of each and are interested in redevelopment. Cytec/Cyanamid has participated in tests, clean-up, and remediation, although the land has not been fully tested or remediated. In contrast, Union Carbide abandoned its Highland Avenue factory without taking responsibility or being held accountable for clean-up costs. Where Cyanamid was a flagship industry and was once the largest employer in the city, Unite Carbide was one of many industries, and the Highland site carries little individual significance to local memory and history.

In contrast with Niagara Falls, Ontario, the high number of illnesses in Niagara Falls, New York, was a well-established fact among many of the people I interviewed, including city officials. One resident spoke about the connection between health risks and chemical industries as something that was obvious in the present but had been neglected in the past:

> The epidemiology of this place is not all that pretty. You had chemical workers dying, you had people working on nuclear programs during the Cold War here; people were dropping like flies. Nobody got the connection. And nobody cared either, because their families relied on plant jobs and the community relied on plant taxes to keep everything moving along. (interview, 23 March 2007)

The issue of verifying the epidemiology of the area did not arise in my Highland Avenue interviews, and for most people, the link between local health problems and the environment was common sense. Cyanamid and Union Carbide represent an opposite politicization of the connections between health problems and proximity to toxic industrial sites. In the case of Cyanamid, corporate and government experts denied any association between health and environment on the basis of a lack of scientific evidence, while workers and residents argued that the connection was dangerous and obvious. In the case of Union Carbide, most people accepted that connection, but some of the workers and residents in the area remained sceptical. Despite the concentration of health problems in Highland Avenue, there was little contestation within the community over the issue of contamination (a topic that will be addressed in chapter 7 through the local politics of community development). More than anything, residents simply wanted to have the abandoned factories in their neighbourhood cleared away.

Stories are All We Have Left

Living memories and experiences of the toxic legacies of industrial ruination in Highland Avenue, Niagara Falls, New York, revealed both traumatic memory and ambivalent nostalgia, paralleling the stories of Niagara Falls, Ontario, although the contrasts were more pronounced. In Ontario, residents and workers accepted that pollution and related health problems were a negative part of an industrial history, and had different views of the industrial history itself, but in New York, this distinction was less clear. People generally saw pollution and health problems as negative, but some felt they were preferable to economic devastation. As in Ontario, people had mixed feelings about sites and processes of industrial ruination, but there was a greater sense of frustration and despair. In particular, the stigma of Love Canal and the dramatic industrial decline in the late 1970s and 1980s were, for some people, entangled. These people blamed city factory closures on the bad press associated with Love Canal and denied the health risks associated with chemical industries: "It was just the attitude, hey, I work here every day. This stuff doesn't hurt you" (interview, resident, 23 March 2007). Other people were more ambivalent: upset about the loss of jobs and prosperity in the city, yet aware of the costs of toxic pollution. A man who grew up in Niagara Falls during the 1970s recalled hearing the warning that "when the smell goes, the jobs go" (interview, 28 March 2007). Finally, some people were worried about the implications of the chemical industries for the health of their communities. In my analysis of these responses, I will again focus on different illustrative narrative accounts that are particularly revealing of living memory in Highland Avenue.

Mary, an elderly African American Highland resident, moved from Alabama to Niagara Falls in the 1950s with her husband, now deceased,

Figure 2.5 Highland Avenue brownfield, Niagara Falls, New York, March 2007

who had found work in a chemical factory (interview, 28 March 2007). Together, they raised ten children and took great pride in the home they had made. Mary was a Baptist minister at one of the eighteen local churches. She said that all of the Highland churches were like one big family, and she was very active in church and community life. She was nostalgic about the golden days of downtown Niagara Falls in the 1950s:

> Downtown was beautiful when I came here, not just beautiful but it was nice. The clothing stores, and there was so much down there that you could enjoy. Theatres, they had three, on Falls Street, Main Street, on Pine Avenue. It was nice. You could go places to sit down, relax, and enjoy. But through the years everything changed, and they took the theatres out from downtown which was a great mistake.

Mary's account echoes those of several elderly white women I spoke with at a seniors centre just up the road from the Highland area. They remembered when the streets downtown were bustling, when cinemas and train stations were open, and when they could walk across the bridge into Canada. They talked about furniture shops, parades, cultural events, theatres, five-and-dime stores, and the wide selection of fashion goods. Most of the women I spoke with at the centre were keen to paint a positive picture of Niagara Falls, "no matter what they say" (group interview, 28 March 2007), and, unlike the men at the centre, avoided talking about the personal impact of factory closures or industrial decline. In contrast with these women, Mary described the devastating impact of job losses and factory closures on families: "How can people survive; how can they live? Your whole livelihood is gone if you close the factories and things. You don't have income. My family, my grandchildren, they are leaving as they are graduating and going to college, they leave the city." For Mary, industrial ruination in Niagara Falls had clearly negative consequences for jobs, families, and the city. Yet she stressed that despite the loss of industry, her friends, her husband, and her health in later years, her own life in Highland had been a good one. She loved being part of a close-knit community where everyone supported each other.

When I interviewed Mary in her home, she was on a respirator. She told me that she had cancer, a bad chronic cough, and no feeling in one arm. Despite her health, Mary was reluctant to draw any connections between community health problems and the factories. She explained her uncertainty: "I can't say it's [health problems] from working in the factories, we just have health problems. I didn't work in a factory. I never worked in a factory." After she described the varying states of health of several people she knew in the area, she added: "There were some that worked in factories, and I don't know if their deaths or sicknesses were related to the factory work that they was doing or not, I can't say that." These statements show that Mary did not consider residential proximity to factories in her contemplation of the causes of health problems in the community. However, she also recalled an instance when she thought something was "a hazard to health": when she lived in a different house in the neighbourhood, there was "black stuff, shiny stuff, and it would get all in your windows, all on your porches. You were constantly cleaning it, all the time." While Mary was aware that the Highland environment could be a threat to health, she was unwilling to blame the factories for local health problems. She was deeply attached to the history and landscape of her home and community. Her account of traumatic memory and nostalgia echoes Ben's Cyanamid account, both in her reluctance to blame anyone for the health problems in the area and in her positive view of the community associated with the heavy industrial era.

Another Highland resident, Dan, had a more extreme perspective about the toxic legacies of Niagara Falls: he argued for a return to heavy industry, even at the cost of community health. His account was deeply nostalgic, and also represented a case of traumatic memory. Dan had lived in the Highland community for his entire life, and was aware of the socio-economic impacts of industrial decline on the city, such as job losses, significant depopulation, and crumbling infrastructure. His father came from the southern United States to Niagara Falls in the 1950s to work in the chemical factories. At the time of our interview, Dan had just been laid off from his job at one of the few remaining plants in the city. He was nostalgic for the days of plentiful jobs, and he felt that a return to industry was the only way forward for the city. He did not see any future in tourism. He described an elaborate plan whereby Niagara Falls could reposition itself as a manufacturing area in a changing economy, particularly when China's economy inevitably declines (as he believed it would) as the United States Rust Belt economy did in the 1970s and 1980s. He argued that metals and other raw materials will always need to be refined, and that there will always be an important role for heavy industry in the global economy. After this argument, he concluded:

> We're uniquely qualified to make it [heavy manufacturing products] here. We could do it cleanly and cheaply, more cheaply because of low-

> cost electricity. We could find a balance between health issues, quality of life, and the environment and the economy. We could find that balance, and I think we could do it right here if we were smart enough to recognize that the future's going to change, and we need to poise ourselves for something other than tourism, that the tourism market is something that people who are far from any natural resource have been able to do far better than we've been able to do so. There's a cost. We have to recognize, *yeah, some people are going to get sick.* We have to look at societies that look at macro-level issues and not just micro-level issues, but we are people who are more into the micro-issues than the macro-issues. (interview, 21 March 2007)

One of the men I spoke with at the seniors centre in Niagara Falls made a similar remark, arguing that the city council should send "two intelligent men to Japan and China and see if they want to invest in Niagara Falls like they did in Canada, and we'll give them cheap power and low taxes" (interview, 11 April 2007). However, Dan's suggestion of a full-scale, Niagara Falls- and America-based return to heavy industry, was more radical. It reflected nostalgia that was not only limited to the industrial past but had been extended into a somewhat fantastic vision of the future.

Carl, another former Highland resident, had point of view that reflected greater ambivalence about the past. In many ways, Carl's story echoed Mary's and Dan's, for his family also moved from the South around the time of World War II, and he grew up in the Highland neighbourhood. However, unlike Mary and Dan, Carl had moved away from the area; for the past thirty years he had been living in a nearby middle-class, largely white neighbourhood, just over the railroad tracks. He pointed out his house on our driving tour, and discussed how his family was the first black family on the block, and how over the years a few others moved into the houses immediately next to his, rather than throughout the community. At the time of our interview, Carl was the manager of the Highland Community Revitalization Committee, a non-profit, charitable organization of citizens and residents committed to promoting community development. Since he moved away from the area, he had a less personal connection to the decline of Highland, was more aware of the health impacts of contamination through his engagement with community development, and was more ambivalent towards the area. He had also experienced racism throughout his years in Niagara Falls, particularly through living in a predominantly white neighbourhood, and he did not see the glory days of heavy industry as any better for race relations. For example, he noted that when World War II ended and the troops returned, black employees were either laid off or demoted to menial positions. Mary's husband, in fact, was laid off from the plant he had moved to work for two years after he had arrived in the city (interview, 28 March 2007). Carl was not nostalgic for the past, but he regretted that "stories" are all that people in the Highland area have to show for years of working in the plants, raising families, and building homes:

> A lot of people do have connections and a lot of pride about working in the plants and helping to build the plants and helping to see production expanded, and you know, maybe even things that they've suggested that went into making the plants more profitable. But you know, it's only, the only thing that they have to show for that, you know, are the stories that they're able to tell and their family members have passed along. (interview, 23 March 2007)

In this sense, the stories can be framed as forms of "living memory," the living collective memory of a community, including its history, people, hopes, expectations, values, conflicts, successes, and failures. Carl's sadness comes from the immateriality of these memories, while in contrast the plants are abandoned, derelict, and full of toxic contamination, and the homes, however well-kept, are situated amidst poverty, contamination, and social exclusion.

For residents of Highland Avenue, Niagara Falls, New York, memories and experiences of industrial ruination related primarily to the trauma of loss – of industry, identity, friends and family, jobs, and health. Residents expressed ambivalent nostalgia for the former industrial era of plentiful jobs, a thriving downtown, and toxic pollution. Yet in the neighbourhood itself, surrounded by railroad lines, crumbling roads, condemned buildings, and abandoned chemical factories, there was nonetheless a strong sense of home, church, and community solidarity for those who remained, a subject that I will explore in greater depth in chapter 6.

Conclusion

As the toxic tours of the 1980s highlighted, there are multiple hidden Love Canals in Niagara Falls that have yet to be officially recognized, both as concrete problems and as lessons about past, present, and future practices. The toxicity of these sites may have dissipated over the years, or it may have lingered and shifted, but it remains a serious issue today. The complex problem of contaminated sites compounds the social and economic issues associated with industrial abandonment: job losses, depopulation, social deprivation in adjacent neighbourhoods, and the precarious nature of employment in the hotels and casinos of Niagara Falls, which are now the biggest employers on both sides of the border. The property market is also at odds with the incentive to investigate sites for clean up, as real estate investors do not want to advertise that contamination could be an issue and both city councils depend on developers and investors. This was one of the biggest barriers in addressing the crisis in Love Canal: residents did not want their new houses to lose their value.

Both cities of Niagara Falls, despite their differences, represent cases of ambivalent nostalgia and traumatic memory in the lived experiences of the toxic legacies of industrial decline, and of the intertwined benefits and costs of chemical industries. In many ways, Dan and Jim's stories represent opposite perspectives. Dan wanted the chemical industries to return in the name of a stable economy, while Jim wanted to clean up the contamination caused by chemical industries in the name of a clean and safe environment. Both were concerned about the future of their respective families and communities. Dan had witnessed the socio-economic devastation of Highland alongside the decline of industry in Niagara Falls. He knew about the health problems associated with heavy industry, but he was willing to risk them as an inevitable cost of economic prosperity. Jim had witnessed deaths and illnesses that he believed were caused by exposure to toxic chemicals, but he had not experienced the severe poverty and social exclusion evident in Highland. Although residents in both case studies reported similar incidents of illnesses and cancer, their stories reveal variations in lived experience of industrial decline. Mary and Ben's accounts reflect similar contradictory feelings in regard to the legacies of industrial decline, particularly the reluctance to ascribe blame and negativity towards companies when their memories and associations with the old industrial era, including concepts of community and fellowship, were positive. Finally, the accounts of older citizens, such as Sharon in Ontario and Mary in New York, reflected different attitudes of older people to industrial legacies. The differences between their accounts suggest that nostalgia for industry is not necessarily connected to generational differences, but relates more to social divisions such as class, ethnicity, and gender.

Another common theme is that of severe health and environmental impacts on residents. Whether these impacts are quantifiable or not, and the extent to which they are real or perceived, is not the topic of this book. Rather, this study points to a sociology of perception of risk, and the stress, fear, and disruption in people's lives in relation to toxic legacies of contamination. Many people have suggested that health problems are clustered in the communities adjacent to industrial sites, which themselves are possible spatial manifestations of discrimination along lines of class and race (see chapter 5). Certainly, socio-economic conditions in both areas indicate spatially defined divisions. There are many differences between the two cases, most notably the higher level of environmental regulation and corporate accountability and the more hidden impacts of industrial decline in Canada, and the widespread and visible impacts in the United States. There is a need for policy to address toxic legacies, to properly investigate the epidemiology of the area, to consider issues of corporate and governmental responsibility, and to find ways to remediate areas. This is no easy task, as Niagara Falls faces the double burden of industrial decline and toxic contamination.

Chapter Three

Protracted Decline and Imminent Regeneration: Memory and Uncertainty in Walker, Newcastle upon Tyne

A twenty-four-metre-high steel angel with a fifty-four-metre wing span stands on top of a grassy hill in Gateshead overlooking the A1 motorway into Newcastle upon Tyne (figure 3.1). The Angel of the North was built over an old coal mine in 1998 by the internationally renowned sculptor Antony Gormley to commemorate the industrial past of the North East of England and to symbolize a shift towards a post-industrial future.[1] This theme connects with the literary and social theorist Walter Benjamin's (1999) concept of the "Angelus Novus," the angel of history who looks back despondently at the ruins of the past as he moves forward with the inevitable progress of time:

> A Klee painting named "Angelus Novus" shows an angel looking as though he is about to move away from something he is fixedly contemplating. His eyes are staring, his mouth is open, his wings are spread. This is how one pictures the angel of history. His face is turned toward the past. Where we perceive a chain of events, he sees one single catastrophe which keeps piling wreckage upon wreckage and hurls it in front of his feet. The angel would like to stay, awaken the dead, and make whole what has been smashed. But a storm is blowing from Paradise; it has got caught in his wings with such violence that the angel can no longer close them. This storm irresistibly propels him into the future to which his back is turned, while the pile of debris before him grows skyward. The storm is what we call progress. (Benjamin 1999, 249)

The passage demonstrates a moment of hesitation, uncertainty, and reflection between moving forwards and looking backwards in time; a profound and violent tension between the past, present, and future.

Benjamin's angel of history reveals a greater sense of conflict in the transition between the past and the future than the Angel of the North, which appears calm and steadfast. The Angelus Novus is male and embodies the gendered foundation of modernity and industrialization. By contrast, the Angel of the North is deliberately androgynous, like many of Gormley's sculptures, and is an appropriate bridge between the male-dominated working-class industrial history of the North East and the increasingly feminized post-industrial labour market. But whether nostalgic or hopeful, male or gender-neutral, both angels stand at a crucial point in history, caught in a painful transition between "destruction" and "creation."

The sense of uncertainty captured in the angels resonates with the contradictions of post-industrial transition embodied within the city of Newcastle upon Tyne. The banks of the Tyne, where shipyards and factories once stood, have been transformed into spaces of leisure and consumption, which include expensive quayside apartments, restaurants, bars, night clubs, art galleries, and museums. The most recent urban renewal efforts in the area have been concentrated on the derelict land of Gateshead Quays, across the Tyne from Newcastle city centre, with the 2002 transformation of the 1950s Baltic Flour Mill into the Baltic Contemporary Art Gallery, the construction of the Millennium Bridge that connects the municipalities of Newcastle and Gateshead over the Tyne, and the 2005 Norman Foster-designed Sage Music Centre with a curved glass exterior. This strategy of arts-led regeneration has been heralded as a good example of post-industrial renewal: "Gateshead Quayside stands as one of the clearest examples in Europe, and perhaps the world, of urban regeneration led by arts and cultural investment" (Bailey et al. 2004, 51). The East Gateshead Partnership, which was established in 1995 to attract property and inward investment and to re-brand Gateshead as a "city of art and culture," spearheaded the revitalization. The projects were largely funded by a £250 million investment from the National Arts Lottery and Millennium Funds. The extent to which these initiatives have helped local people in Newcastle and Gateshead in terms of job creation, overall economic growth, and historical and cultural identity is not yet known.

Beyond the sites of regeneration on both sides of the river, there are considerable areas of dereliction. Cameron and Coaffee (2004, 11) highlight the contrast between regeneration and decline as follows: "The River Tyne cuts between Newcastle and Gateshead in a steep-sided valley, with the new developments below its slopes. Linked and enclosed

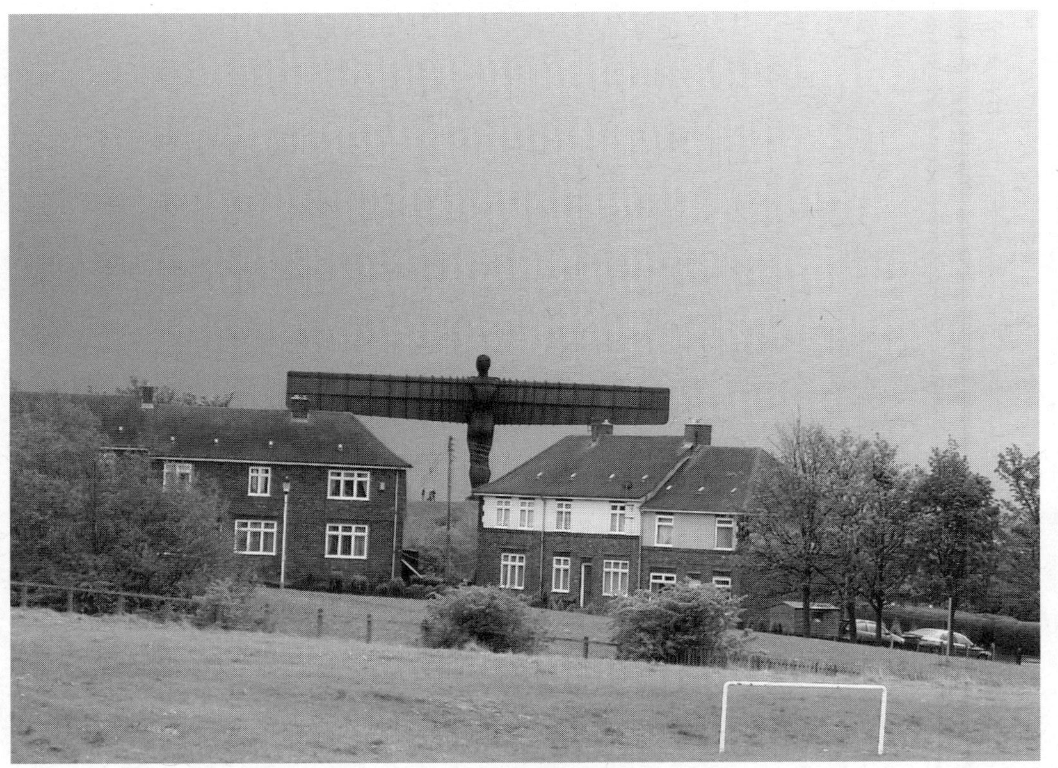

Figure 3.1 Angel of the North, Gateshead, North East England, June 2005

by the bridges across the river, the developments now on either side form a veritable amphitheatre of urban renaissance, but one which appears to be physically divorced from the urban areas beyond the slopes." Similarly, Robinson (2002, 320) describes spatial contradictions between areas of regeneration and areas of deprivation as part of the overall fragmentation and diversification of the North East region. He notes that a visiting journalist who wanted to report on the North East's "renaissance" could easily cite the MetroCentre shopping complex in Gateshead and the Quayside amphitheatre attractions as sufficient evidence of urban renewal. At the same time, Robinson argues, a journalist could just as easily write an article on the North East as a socially and economically depressed place, if she were to look beneath the surface.

The focus of my research in Newcastle was on one such space beneath the surface: the Walker community, located in the East End of Newcastle. Walker, a traditional working-class neighbourhood, has long been associated with the shipbuilding industry on the River Tyne, from the bygone days of bustling shipyards and cranes to the present state of industrial decline. The local shipbuilding industry declined due to the loss of global competitive advantage within the British shipbuilding industry as a whole during the first half of the twentieth century. Britain had dominated the international shipbuilding industry in the late nineteenth and early twentieth centuries, producing 60 per cent of the world's ships as late as 1913. By the 1960s, that number fell to less than 10 per cent. According to Lorenz (1991), the decline of British shipbuilding related to management's failure to shift from "craft" to "bureaucratic" methods of mass production in the two decades following World War II. The British shipbuilding industry was based on labour-intensive craft skill, which had been a competitive advantage during the pre-war era when there was fluctuating demand for specialized, non-standard vessels (Lorenz 1991). However, in the post-war era, shipyards in Sweden, Japan, and West Germany gained a competitive advantage through their model of mechanized, standardized shipyards with fewer skilled labourers, which suited the new shipbuilding standards of international vessels. Consistent with the pattern of the British shipbuilding industry, most of the shipyards on the Tyne closed between the 1960s and 1980s. The iconic Swan Hunter shipyard was bolstered by lifelines for building government warships and dismantling ships before finally closing in July 2009. Most United Kingdom shipbuilding companies were acquired by companies in other parts of the world, or began to operate in other sectors, such as engineering, ship repair, and ship conversion, in other parts of the country.

The landscape of abandoned shipyards, derelict warehouses, and under-used properties attests to Walker's legacy of industrial ruination. Among the abandoned shipyards and warehouses of Walker Riverside, I observed new offshore oil and gas companies, scrapyards, and the remains of Swan Hunter, the "last shipyard on the Tyne." The co-presence of these new, old, and dying industries reflected the slowness and difficulty of socio-economic change as spaces caught between the destruction and creation of capitalist redevelopment.

The period of my research in Walker, from June 2005 to March 2006, represents a crucial point in the experience of industrial ruination as a lived process, just before an ambitious regeneration scheme was fully underway and just before Swan Hunter closed in July 2006. When I started my research, Newcastle City Council and its regeneration development partner were still in the planning and community consultation phases, the latter of which was a necessity, given the community's resistance to the proposed plan. The regeneration process began as I finished my research, with the first wave of demolitions in progress and the first replacement homes under construction. The local residents and workers I spoke with were caught within a long process of change, and were uncertain as to what the future would bring. Both agents of regeneration and members of the community itself were eroding the connection between Walker and shipbuilding. For many people in Walker, shipbuilding represented something that had been important in the past, but was less important than the impending regeneration of their community and the potential demolition of their homes. Many people identified with the old industrial sites and memory of shipbuilding only through the notion of regeneration. Yet Walker has remained a place with a strong tradition of close families and community networks, much of which is based on a shared industrial history.

This chapter explores the landscapes and legacies of industrial ruination and urban decline in Walker. As I have mentioned, at the time of my field research, the prolonged decay of the shipbuilding industry had not resulted in its full demise, and this lack of a clear break with the industrial past has had important implications for collective memory. In Walker, memories of the industrial past have yet to be officially or unofficially reconstructed. Local accounts of sites and processes of industrial ruination represent living memories that are defined by a lack of closure with an industrial past. These living memories are situated in the context of "industrial ruination as a lived process" and, unlike simple nostalgia or commemoration, they have emerged in diffuse ways. This chapter argues that collective memory, defined through the social reconstruction of the past, remains uncertain in cases such as Walker, where industrial decline has been protracted, traces of old industrial activity remain, and regeneration has yet to transform the landscape of industrial ruination.

Walker, Newcastle upon Tyne

My research focused on the relationship between the residential community of Walker and the Walker Riverside industrial area along the River Tyne. Walker has steadily declined over the past forty years, with the erosion of housing, shops, and services accompanying the industrial collapse. The City of Newcastle as a whole experienced significant depopulation between 1971 and 2001, but the loss was more severe in Walker, which dropped from a population of 13,035 in 1971 to 7,725 in 2001. Walker ranked the worst of all twenty-six wards in Newcastle and thirtieth worst of all 8,414 wards in England against the 2000 English Indices of Deprivation (Noble et al. 2000). The area also has rates of cancer, heart disease, and respiratory illnesses that are significantly higher than the rest of the city (MacDonald 2005). Various sources, including interviews, newspaper articles, and city council documents, indicate that there may be some residual contamination in parts of Walker from industrial activities, and newspaper articles in the Newcastle *Evening Chronicle* have highlighted deaths of former shipyard workers from cancer caused by exposure to asbestos (see Sadler 2006). There are parallels between the asbestos cases and the health effects of toxic contamination in Niagara Falls, although none of the residents, workers, or former workers I spoke to reported health problems related to the shipbuilding industry.

Walker is located roughly five kilometres east of Newcastle Central Railway Station, and comprises the area between Welbeck Road and the River Tyne. Local people often speak of Walker as extending beyond the boundaries of the Walker ward (as defined by the city council) to include the neighbouring areas of Daisy Hill, Eastfield, Walkergate, Walkerdene, and Walkerville. Walker Riverside refers to the area of Walker which is closest to the River Tyne, but the boundary between Walker and Walker Riverside is also blurry (Madanipour and Bevan 1999). Madanipour and Bevan (1999) suggest that socio-economic differentiation within Walker follows topography. The riverbank, the lowest part of the community, is the poorest area, whereas uphill, north of the river towards Walkergate, there are higher income levels. The authors' topographical characterization of the area reflects patterns of economic growth and decline: communities formed around industries along the river, and as those industries declined, so too did the communities, particularly those closest to them.

At the time of my fieldwork between June 2005 and March 2006, 70 per cent of the total housing stock in Walker consisted of local authority housing, most of which was built during the 1930s. The East End of Newcastle developed in the late nineteenth century to house industrial workers in shipbuilding, engineering, coal mining, iron, and chemical and glass works. Houses in Walker were built to accommodate people who worked in the shipyards along the riverside, and so the flow between the community and the industrial riverfront was initially integral. That flow has since been disrupted, as the Walker community is now physically separated from the industrial riverside area by a major road, security gates, and fences (figure 3.2). There were fierce debates at the time of my research over the quality and value of this local authority housing stock. The proponents of city council-led area regeneration viewed these houses as "stigmatized" and unsightly, while the residents viewed their homes as sturdy, well-maintained, of good quality, and layered with family histories and memories.

Traditionally, Walker is a predominantly white, working-class area, like the North East and the city of Newcastle upon Tyne more widely. At the time of the 2001 Census, 97.2 per cent of the population in Walker was white, in contrast to 93.1 per cent of the population in Newcastle and 92.1 per cent in the United Kingdom. In an ethnography of youth subcultures in North East England, Anoop Nayak (2003) analyses "Geordie" (Newcastle) identity through the lens of class and race. Nayak argues that it is a fallacy to presume that racism is absent from predominantly white areas, and traces complex histories and trends in race relations in the North East. Although racism may have long historical roots in the region, city, and community, the issue of racism and difference has become increasingly integral to an analysis of the Walker community since the 2001 Census (results from the 2011 Census will be released in 2012). In 2000, the United Kingdom government introduced a dispersal system whereby asylum seekers were to be placed in designated deprived areas in "hard to let" housing. Asylum seekers were "dispersed" throughout the North East, and a relatively high number were placed in Walker. Neither the local population nor the community centres in Walker were prepared for this sudden influx of people. Moreover, as a deprived community in its own right, many residents were resentful that asylum seekers were provided with community housing and other resources, however meagre. The asylum seekers received a basic package, including basic accommodation and an allowance amounting to 70 per cent of income support at £40 a week for a single person (interview with a charity for asylum seekers, 1 December 2005). In 2003, asylum seekers lost the right to apply for work, although limited permission to do so was granted in 2005. At the time of my research, there were approximately 400 asylum seekers in the Walker and Byker area. The dynamic between the primarily white, working-class community of Walker and the primarily black African asylum seekers who have arrived in Walker since 2000 sheds light on elements of the community as a whole. These issues will be discussed later in this chapter, in relation to "community solidarity" as a legacy of industrial ruination.

Long-term residents in Walker have lived through four decades of slow and protracted decline in the shipbuilding industries along the river. They have seen many attempts at regeneration over the years, with the earliest plans dating from the 1980s, but nothing has reversed the trends of depopulation, stigmatization, and deepening social and economic deprivation in the area. Walker has been the target of Newcastle City Council regeneration efforts since 2001, which have the stated aim of attracting new residents and reversing the retreat of public and private services. In 2002, the city council–led Walker Riverside regeneration plan became part of the Bridging Newcastle Gateshead Housing Market Renewal (HMR) Pathfinder, one of nine HMR areas identified in the North and Midlands of the United Kingdom as areas of "low housing demand" with depopulation, dereliction, poor services, and poor social conditions (Long and Wilson 2011). Following trends across similar deprived communities in the country, the Walker regeneration plan was based on the demolition of over 700 homes and the construction of apartments to attract middle-class populations in their place. Most of the homes targeted for demolition were located on the riverside, which is considered key potential real estate. While a similar process of regeneration and relocation occurred in Newcastle's West End (Robinson 2005), Walker was a more difficult case because of residents' strong attachment to their community. Most people in Walker opposed the demolitions proposed within the city council's vision of change for the community and protested the regeneration scheme. In 2006, after a contested series of "community consultations" and "master

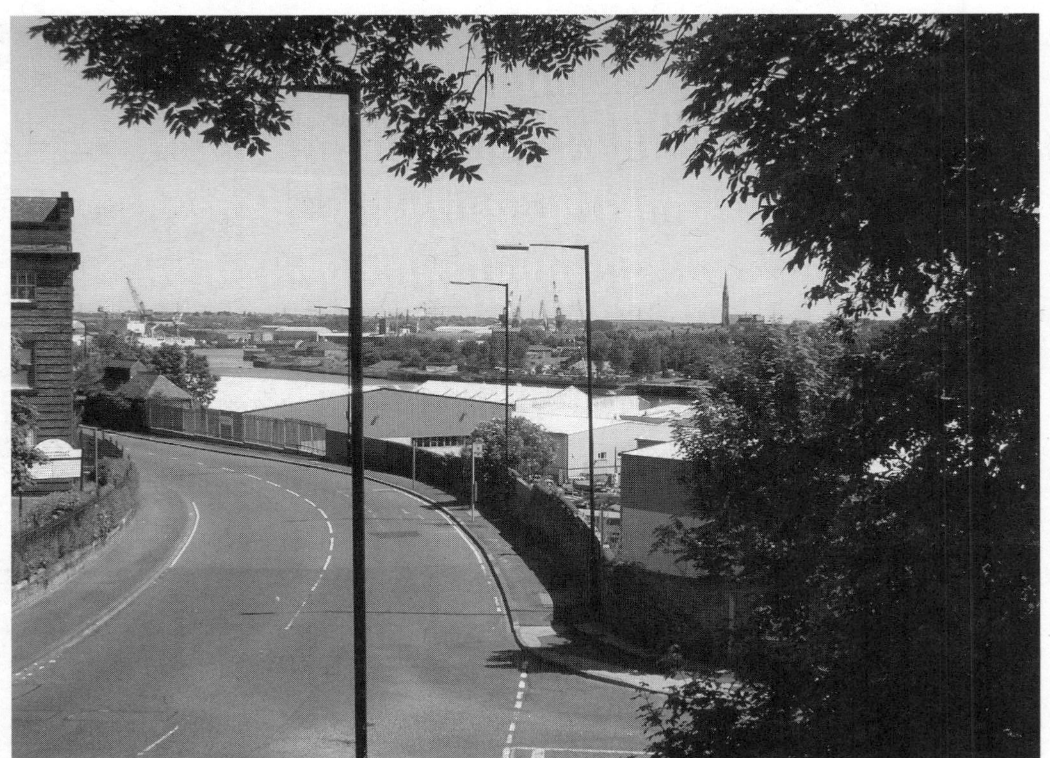

Figure 3.2 Road between Walker community and riverside, June 2005

plans," the city council and its regeneration development partner finally gained enough approval to proceed with their plans under "option three" with "major impact" for the community, which involved maximum housing demolitions and developing shops and services. The six-year period of imminent regeneration in Walker left people profoundly uncertain about the future of their homes and their community, as they had no guarantee of what the post-industrial future would hold. When one resident learned that her house was going to be demolished, she experienced extreme mental anguish. She described her experience of uncertainty in the face of change as incapacitating, removing her ability to speak, move, or act:

> Now they're saying it's [my house] going to be demolished, and I just hit rock bottom, I just, nothing, I couldn't get past my front door, I could hardly speak. So, there's people like me that have panic attacks, and they're just not going to listen to us. These things are cut and dried before you even get to know. I mean, Bernard Street, the plans were there ten years before they even told the residents, before they even let it leak out that they planned on demolitions. (interview, 12 September 2005)

Industrial Ruination in Walker

The physical landscape of industrial ruination in Walker is concentrated along the seventy-hectare Walker Riverside industrial area, once the heart of shipbuilding on the Tyne. This area has been industrialized for over 150 years, and has been used for several industries, including lead, iron, copper, chemicals, alkali, tar works, coal mining, stone quarrying, and shipbuilding (MacDonald 2005). Local residents also remember tobacco industries and a glue factory (interview with three residents, 20 March 2006). At the time of my research, the land was used for marine and offshore industries, scrap yards, ship decommissioning, and limited shipyard activities. Despite the devastating impact of Thatcher's policies for shipbuilding in Newcastle and manufacturing in Britain more generally, both the city council and the central government have tried to keep the iconic shipbuilding industry afloat – if not economically, at least symbolically – since it went into decline. Lifelines and contracts have been offered to Swan Hunter shipyard over the years in an attempt to forestall the closure of the "last shipyard on the Tyne." The end of the shipbuilding era was not properly marked until Swan Hunter closed in 2006.

One of the most defining characteristics of the Walker Riverside industrial area is its physical separation from the Walker residential community. Walker itself is also socially, economically, and geographically separated from the rest of Newcastle because of poor public transport; stigmatization as a deprived area; lack of shops and services; high poverty, crime, and drug and alcohol use; and many abandoned or degraded properties. Another defining characteristic of the industrial area is that, unlike many other old industrial sites in the North East, Walker Riverside has retained its industrial designation: the area is a mixture of abandoned, partially working, and fully operational sites. Newcastle City Council owns the majority of this area, and regulates and restricts its industrial and non-industrial uses. However, the security gates around the abandoned shipyards and warehouses have not prevented activities such as vandalism, arson, the climbing of cranes, scavenging, drug and alcohol use, and general hooliganism.

Tim Edensor (2005) examines meanings and uses of British industrial ruins in his book *Industrial Ruins*, which catalogues some of the "contemporary uses of industrial ruins," from plundering to home-making, adventurous play, leisure, and art space. Walker Riverside fits into this typology to some extent, although its spaces are managed and restricted, and thus the activities that occur within them are either employment-related or illegal. This is in stark contrast to old industrial sites in cities such as Berlin, Manchester, and New York, which have often been the sites of artistic, cultural, and political activities. One major cultural activity was organized at the near-abandoned Swan Hunter shipyard in May 2006: a Pet Shop Boys rock concert against a projection of the cult movie *Battleship Potemkin*, as part of the Newcastle Gateshead initiative to promote the city (McMillan 2006). However, this was a regulated, rather than spontaneous or subversive, cultural activity, and it fit with the "competitive city" and cultural re-branding approach of the city council.

Site observations of the Walker Riverside industrial area gave me some sense of the scale of the former industries and their historical economic and symbolic importance. I visited the iconic Swan Hunter in December 2005, seven months before its final closure (figure 3.3). Swan Hunter is located at the far northeast end of Walker Riverside industrial area, technically within the ward boundaries of Wallsend, North Tyneside. I had arranged to meet with the shop steward and representative of the General, Municipal, and Boilermakers Union (GMB) at 8:30 a.m., the start of the work day. I waited for three quarters of an hour in the Swan Hunter security room, where workers in hard hats and vests with fluorescent stripes flowed in through the glass front doors and out through the back doors into the shipyard. People in suits with briefcases also streamed in through the front doors, but passed through a side door into the corporate offices. Sitting in the front entrance room to the last shipyard of the Tyne, months away from its final closure, felt like stepping back in time.

By the time I visited the famous shipyard, the story of Swan Hunter had already been well-documented by local historians and journalists. As Peter, the senior payroll manager at Swan Hunter, noted, "I know it's for your project or whatever, but it has been documented many times and it's in many books and whatever. History, if you like. It's of great interest and in years to come it'll be even greater, especially when this place is gone" (interview, 2 December 2005). Local books such as *Lost Shipyards of the Tyne* (French and Smith 2004) and *Swan Hunter: The Pride and the Tears* (Rae and Smith 2001) tell some of these tales. However, these stories are not complete, as the processes of decline are still underway. Rae and Smith's (2001) story of Swan Hunter has a happy ending, of a "phoenix rising from the ashes" via the 1995 rescue of the shipyard from closure. The story of near-death and resurrection through lifelines continued beyond 1995; Swan Hunter was awarded the first ever shipbreaking licence in May 2006: "ironically, the bid to save the yard, which has built some of the world's finest ever seafaring vessels, could now see Swan's workers destroying some of their own ships" (Whitten 2006a). It finally closed in July 2006, which resulted in 260 job losses. A task force in North Tyneside was subsequently established to decide what to do with the yard. Included among the North Tyneside City Council plans for the yard was the idea of turning it into a maritime museum that would both celebrate and mourn the shipbuilding history of the Tyne, and this was just one of many possible futures. After years of struggle, near-death, and rebirth as the last shipyard of the Tyne, Swan Hunter shipyard is no more. Its floating dock, equipment, and cranes were sold to Bharati Shipyards, India's second-largest private shipbuilder, in April 2007 (Reuters 2007). The last two cranes of the shipyard were detonated by explosives in June 2010 (Fletcher 2010). Like the Angel of the North, Swan Hunter remains a symbol of painful post-industrial transformation.

One of the most visually impressive sites of ruination that I encountered was the abandoned A&P Tyne Shipyard (figure 3.4), located just southwest of Swan Hunter along Walker Riverside. The shipyard had been stripped completely of cranes and equipment in what one city

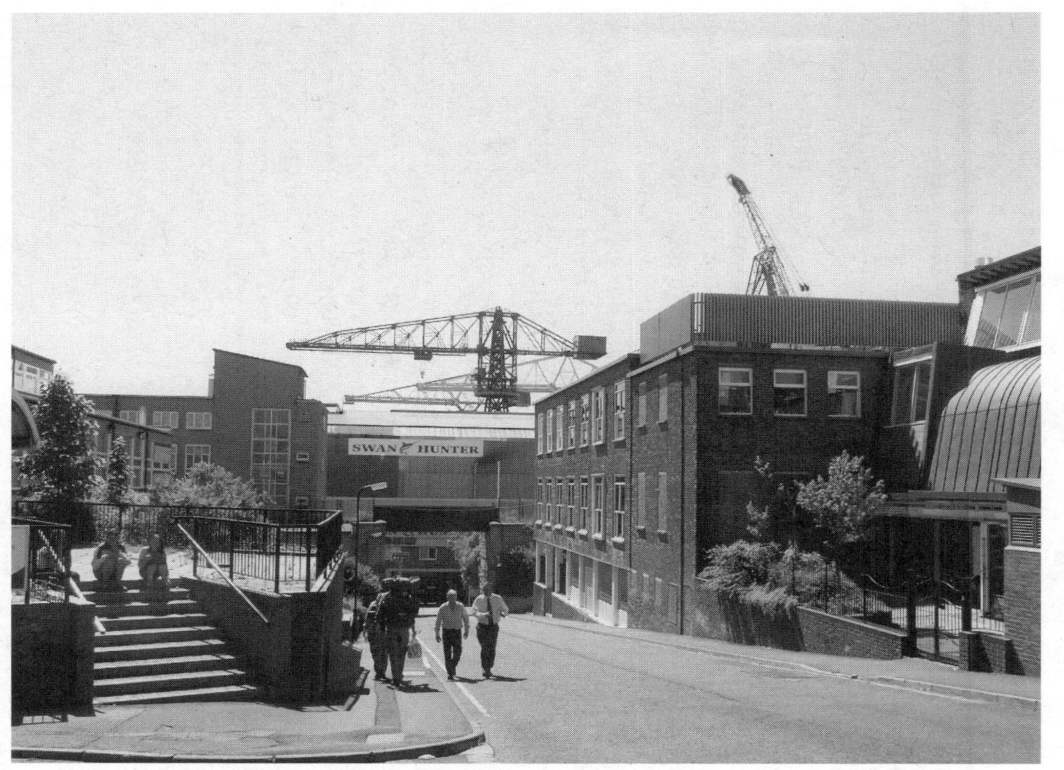

Figure 3.3 Swan Hunter shipyard, Walker Riverside, June 2005

officer referred to as a "scorched earth policy" (interview, 29 September 2005). The company subsequently refocused its ship decommissioning activities at another site further along the Tyne, where Polish workers staged a "wildcat strike" in May 2006 over poor wages (Whitten 2006b). The physical landscape of this site rapidly changed over a one-year period, from an active ship decommissioning yard, to a youth's playground of disused cranes, to a barren lot. Articles in *Evening Chronicle* described incidents of young hooligans in the disused A&P Tyne shipyard, including frequent arson attacks and the scaling of cranes (e.g., Kennedy 2006). The story of A&P Tyne in 2006 – of "scorched earth" abandonment, relocation, a new and disgruntled labour force, and vandalism – reveals the changing dynamics of capital and labour in the North East. In this sense, one can read this landscape as part of a process of industrial ruination and re-creation over time. As the rest of this chapter will argue, the stories embedded in the material landscapes of industrial ruination also connect with the stories of the people who inhabit such landscapes.

Living Memories of Industrial Ruination in Walker

The concept of living memory suggests that local memories exist within the present as dynamic and changing processes, and that they do not necessarily function as part of the social construction of official or unofficial collective memory. This concept has parallels with Nora's (1989) notion of "true memory" – memory that has not yet been absorbed by official history. Nora introduces the concept of "lieux de mémoire" (sites of memory) to subvert what he describes as the separation between memory and history. However, Nora's lieux de mémoire represent a somewhat idealized vision of memory: they are "mixed, hybrid, mutant, bound intimately with life and death; enveloped in a Möbius strip of the collective and the individual, the sacred and the profane, the immutable and the mobile" (1989, 19), and to be created, they require a "will to remember" and a capacity for change. By contrast, living memory in the context of Walker comprises neither a collective willingness to remember or to forget, nor a full capacity to accept change. Rather, it is rooted in complex and varied lived experiences of processes of industrial decline and post-industrial change. More specifically, it is complicated by the fact that at the time of my research, a single iconic shipyard remained open, yet the shipyards had been in significant decline for over forty years.

Zukin's (1991) concept of an "inner landscape of creative destruction," an interior world of rupture reflected through "individual perceptions of

Figure 3.4 A&P Tyne abandoned shipyard, Walker Riverside, September 2005

structural displacement," provides further insights for situating an analysis of memory in relation to lived processes of industrial ruination. Zukin discusses the ways in which people's interior worlds relate to the socio-economic context of a shift from "landscapes of devastation" towards "landscapes of consumption." As she argues, "deindustrialization and gentrification are two sides of the same process of landscape formation: a distancing from basic production spaces and a movement towards spaces of consumption" (269). "Liminal" spaces, or culturally mediated no-man's-lands, are created in these socio-spatial shifts. Walker, caught somewhere in the middle of a social-spatial shift, is one such space. My analysis of the "inner landscape" of industrial decline in Walker explores a relationship between memory and place that is rooted in the lived experience of memory in the present. According to Savage (2003, 237): "The deindustrialized landscape, like a ruined battlefield that heals over, is ripe for commemoration. As the physical traces of the industrial age – the factories, the immigrant enclaves that served them, the foul air – disappear, the urge to reaffirm or celebrate the industrial past seems to grow stronger." One is left with two questions: (1) what happens to memory when the ruined battlefield has not healed over, and (2) what happens when the physical traces of the industrial age linger? These are key questions to consider in relation to living memory in Walker, and indeed for all three of the cases in this book. However, in Walker, the lack of a break with the past is perhaps more surprising, given the context of long-term decline, imminent regeneration, and post-industrial rhetoric throughout the wider city-region.

The following analysis draws on selected interviews which are particularly illustrative of lived experiences and memories of industrial ruination in Walker. First, I will explore two themes of living memory in relation to industrial ruination, reflecting differences in class and age; and second, I will explore two themes of living memory at the level of the community as a whole.

Where are the Cathedrals of the Working Classes?

Local memory of the shipyards along Walker Riverside stretched beyond the place itself to include Newcastle and the whole of the North East. The interviewees who had the strongest sense of sadness, loss, and disappointment over the decline of the shipyards in the past thirty to forty years were those connected with the industry, either as workers or as close family of workers. I spoke with Roger, a local resident of the neighbouring ward Byker, who used to work in shipbuilding but who now worked for a local charity for asylum seekers in local hard-to-let homes. As he recalled his job interview for work in the shipyards, he recounted with irony how the employers had said he was lucky to get in and that he could have a job for life. He expressed a deep sense of loss connected with the decline of shipbuilding, and remarked:

> Shipbuilding is not like coal mining, unhealthy or unsafe, when there's both a sadness and a gladness when it goes. With shipbuilding, there is just a sadness, a loss, and there is nothing to replace it ... There is a psyche of the North East which is built on the pride of shipbuilding. (interview, 1 December 2005)

His comment about "sadness and gladness" regarding the decline of an industry draws attention to an important distinction between heavily polluting industries and cleaner industries. Although some shipyard workers in Walker have suffered from the effects of asbestos and dangerous physical work in the shipyards, even more dangerous and "dirty" industries, such as coal mining, tend to have more conflicted associations.

One of the few remaining shipyard workers in the North East expressed similar views. Mark, the shop steward at GMB, had worked in the shipyards for twenty years, and had spent the last few at Swan Hunter. He noted that Swan Hunter was a powerful symbol as the last shipyard of the Tyne, and for this reason, people demonstrated a loyalty to the shipyard itself, if not a loyalty to the company. He emphasized the importance of the shipbuilding industry in particular for Britain, asking: "How can an island exist without ships?" These views were also highly politicized, as he highlighted in the following statement:

> Yes, it's of tremendous symbolic and economic importance, and if it goes down under a Labour government we can only conclude that we're not valued as we think we should be. It has been said by elements of government, well let shipbuilding go, the North East doesn't need it, but it certainly does need it. You've got a thousand people working here; well, under a thousand work here, but you've got a vast network of subcontractors and suppliers. And if this type of work isn't here, we won't go and find something else to do; we'll just go and work somewhere else which would be a loss to the area and the local economy. We can't all sell each other baskets and jam. (interview, 2 December 2005)

Mark repeated this last phrase again later in the interview. His narrative of socio-economic change reflected not only deep scepticism about the merits of the service-based economy, but also injustice and anger over the politics surrounding the decline of manufacturing in the United Kingdom. He strongly identified with the politics between Labour and the Tories, and he suggested that it would be bad for Labour's image in the North East if the shipyard closed under a Labour government. To conclude, he lamented: "There's Durham Cathedral, but where are the cathedrals of the working classes? Our cathedrals were destroyed." Then, highlighting his awareness of being interviewed by a sociologist, he said, "There's plenty more where that came from, if only there was time." He had to get back to work. However, for another perspective, he asked me to speak with another man, Peter, the senior payroll manager, who had worked for Swan Hunter since 1969.

Peter gained some local fame as the "last man" and "first man" at Swan Hunter when the company went into receivership in 1994, closed on Christmas Eve day in December 1995, and was resurrected under a Dutch company's ownership in 1996. He was the last person on the books for the yard when it closed.[2] Within the local community, he became a personal symbol of the near-death and resurrection of the shipyard. At the time of our interview (2 December 2005), Peter noted that a feeling of uncertainty was present in the shipyard since contracts were coming to an end and there was not a great deal of hope for securing future work. However, in contrast to Mark, Peter was optimistic about his future. He acknowledged that when the next vessel was finished in a few months' time, he might have to look for another job, which was difficult given his age. He said that he might only get part-time work, and he didn't know what to expect. However, he also said that he was in a fortunate position because he no longer had a mortgage and could afford to "potter" until his retirement. Although he expressed sadness about the passing of the shipbuilding era, he was less angry than Mark, a fact which may have been related to his position as a white-collar, rather than blue-collar, worker within the firm. Furthermore, Peter was a payroll manager, which most companies need, whereas Mark was a shipbuilder, an occupation that is much less transferable.

There were narratives of sadness, loss, and pride associated with shipbuilding in the stories of the relatively few people with remaining direct connections to the industry. These views represent a politics of living memory, as they connect with the wider context of the decline of shipbuilding in the North East and with Thatcherism and the decline of manufacturing in Britain more generally. They also suggest that the sense of loss in Walker is more strongly associated with occupation, class, and gender, as the shipbuilding industry included predominantly male, white, working-class workers (Dougan 1968; Roberts 2007). As Mark provocatively suggested, there are no longer any working-class cathedrals. However, I had the sense that cathedrals of working class memory, in the form of museums, art galleries, or monuments, were not what Mark was referring to. He was referring to cathedrals of living industry rather than remembered industry, such as factories, shipyards, and flourishing industrial communities: cathedrals that have been destroyed.

Generational Nostalgia?

As direct experiences of working in the shipyards have faded into the past, generational differences have emerged in memories and perceptions of industrial ruination. Gary, the regeneration manager for the East End Partnership, described these differences in his interview:

> I think probably older, more senior members of the community have a nostalgic view of activity on the Tyne some years ago and I think their attachment is based around a remembrance of that high level of activity and employment, although quite often that may well be through rose-tinted spectacles, as it wasn't easy, that physical kind of labour. (12 September 2005)

Gary went on to contrast older people's attachment to old industrial sites as romantic and nostalgic with younger people's view of ruins as "playgrounds" or "eyesores." By playgrounds, he alluded to young people's informal uses of space, which included arson, vandalism, climbing cranes, and alcohol and drug use. However, from other interviews, I found that not all older people were as nostalgic.

I spoke with an elderly woman and two upper-middle-aged women at a community centre in Walker, and they demonstrated little attachment to old industrial sites. All three agreed that there was not much activity at the shipyards any longer (group interview, 22 March 2006). When I asked about the history of shipbuilding work in the area, and whether people remained connected to that history in subsequent generations, the interviewees responded by saying that all the people who worked (or who had worked) in shipbuilding were elderly, and that it was up to the younger generation to find alternative employment. Even the elderly woman, Ethel, who was in her seventies, did not associate the shipyards with her own generation; her uncles were her closest relations to have worked in the shipyards. This illustrates that for many residents, shipbuilding long ceased to have relevance as a source of employment for the local community, despite the fact that it continued in a skeletal form until recently.

During an informal discussion with Sheryl, a resident of Walker, and two of her family members, we explored intergenerational memories and perceptions of the shipyards. I accompanied Sheryl to her father's taxi company in Byker. Her grandfather used to work in the shipyards, and she thought that her aunt and uncle, who were working at the taxi company on that day, might be able to tell me something about the shipyards. Sheryl herself had no direct connection with the shipyards, although as a local resident with family links to the industry and social ties within the community she respected the industrial past and the stories of her parents and grandparents. The aunt and uncle recalled that there was once a "bone yard" along the riverside (a glue factory), and they described, with a lively sense of humour, how much it stank. Throughout the discussion, the three relatives would often break into laughter, as they seemed to find the ways of the past funny and odd. This detachment seemed in part to relate to the passing of time, as they were further from the experience of the shipyards, but it also seemed connected to their economic situation, which, because of the taxi business, seemed relatively stable.

Although there were generational differences in local memories and perceptions of the landscapes and processes of industrial ruination, they did not occur in a linear fashion in which the oldest generation had the greatest nostalgia and the younger generations were more detached. People from the older generations also expressed detachment from the history of shipbuilding, even if they were related to people who had worked in the yards. Other factors, such as a direct socio-economic relationship to the industries, seemed to be at least as, if not more, significant. This echoes the findings in the narrative interviews in Niagara Falls. The prolonged industrial decline, which began as early as the 1960s, also accounts for a general acceptance within the Walker community that shipbuilding is an increasingly distant source of pride, and that it no longer has socio-economic meaning for the present. However, although generational, socio-economic, and gender differences could be traced through different individual lived experiences and memories of industrial decline, memory at the level of the collective or the community was consistent across two diffuse forms of living memory: regeneration and community solidarity.

Regeneration

The most immediate issue facing the Walker community at the time of my research was the imminent regeneration that involved the large-scale demolition of homes in which people had lived in all their lives. Most residents saw the decline of shipbuilding and the old industrial riverside as part of that story, but as a non-urgent, inevitable part. By contrast, the prospect of housing-led regeneration was seen by many residents as an immediate threat to their sense of community. People who did not live in or come from the East End of Newcastle viewed Walker as a stigmatized area that had suffered from decline and deprivation for decades. There had been plans to improve the area during that time, but the situation had grown worse rather than better. Residents were sceptical and distrustful of the change that might occur this time around (various interviews, 2005–6). Many felt that regeneration would not serve the interests of the community, but rather those of developers and "yuppies." Their suspicions were well-founded: the planned regeneration involved the demolition of existing homes, the construction of new riverside flats meant to attract higher income groups to the area, and the creation of a "riverside village" concept of shops and services to cater to the new middle-class population. In other words, the proposed physical regeneration was designed to facilitate gentrification.

The issue of impending regeneration tended to eclipse many residents' attachment to the shipyard sites. Apart from those who remembered or sustained direct employment connections with the shipyard industries, sadness, loss, injury, and despair were no longer connected with the memory of those sites, but had been transferred to the present community infrastructure changes. Ethel, for example, had lived in one of the homes set to be demolished for thirty-one years. When I asked about her relationship to, and perception of, the shipyards, she replied:

> My uncles used to work there, but they worked there a lot of years ago, and I've got three sons that work down in Shepherd Offshore. Apart from that there's only one I know. I mean there's not even much going on down there now in the yards; that's slowly going downhill and all. I mean, there's not much I can say really, 'cause I mean at the end of the day there's not much I can do about it now. It's a done deal. People there won't let me know when, how long's it gonna be, before they're coming. But they said they're supposed to build houses on Pottery Bank before they pull them down and I'm hoping they're going to keep that promise. (interview, 22 March 2006)

Ethel began by mentioning her uncles, who used to work in the shipyards, and her son, who worked at Shepherd Offshore, the local offshore oil and gas company located among the former Walker Riverside shipyards. Then she emphasized the declining importance of the shipyards today and quickly returned to the subject of regeneration and the demolition of her home. She described her powerlessness to change developments in the area, since "it's a done deal." I encountered a similar situation when I visited other Pottery Bank residents. Even when I asked questions directly about the shipyards and their decline, it was difficult to get the residents to talk about anything outside of regeneration. They were clearly emotional and passionate about this topic and talked about how the uncertainty of the future of their homes impacted their health and had led to stress, anxiety, poor mental health, panic attacks, and a feeling of inability to do anything with their homes – sell them, re-paint them, furnish them, or anything at all (interview with two residents, 12 September 2005). Once the regeneration and demolitions were underway, the Walker Riverside Regeneration Project team started to brand the area as a "location of choice" and their 2007 Heart of Walker campaign emphasized building on community spirit in the regeneration project, but failed to explicitly mention industrial decline. However, one might argue that "community spirit" in this context is a diffuse reference to Walker's shared industrial past.

Community Solidarity

If the shipyards have gradually lost meaning for some people in Walker, there has been one thing which has continued to have deep cultural and social significance: the community. The sense of community and the collective awareness of community activities is a theme that also connects with local living memory. Memory, after all, is not just about the past, but reflects the present, and particularly how the past filters into the present. The resilient character of the community can, in part, be explained by the fact that its population consists of people who have remained in the area and persevered through economic difficulties. A number of families and individuals have left Walker in the past decades, so the people who remain are either those who have chosen to remain through attachment to their family and community, or else those who have been unable to move due to financial or personal constraints.

In their classic study of the white working-class community of Bethnal Green in the East End of London, Young and Willmott (1957) found strong patterns of extended family networks and solidarity, with grandmothers, daughters, sons, uncles, and aunts all living within blocks of one another and relying on one another for support. I found similar patterns of extended family networks and solidarity at the time of my field research in Walker. Young and Willmott focused primarily on the family dynamics of one community, but they also alluded to the industrial heritage that had shaped that community, built on male-dominated manufacturing work in the factories and casual labour on the docks. In contrast with Young and Willmott's study, I have focused more on the importance of industrial heritage – specifically, the collective memory of a shared industrial past – in shaping notions and structures of working class community even when most of that industry has gone.

Many interviewees referred to the strong sense of community in Walker. A representative from the regeneration partner for Newcastle City Council puzzled over the discrepancy between the physical infrastructure and this community spirit: "There is nothing physically appealing about the area, no landmark buildings and indistinct housing and roads, yet there is a strong local commitment to the area based on its industrial past" (interview, 2 December 2005). A Labour city councillor for Walker who opposed regeneration also referred to the strength of family connections in the community, and noted that families often live on the same street for generations, and that local people, regardless of age, tend not to travel. One could "look at it as a form of stagnation," he said, but also as "a strength and an asset" (interview, 30 August 2005). Three local residents described Walker as a place of good families and community spirit despite being "run down." One resident, Susan, described her attachment to the community:

> Yes, it's run down and everything. But there is a strong community spirit and if anything happens to anybody in the community they'll all rally around to make sure they're alright ... And we know the people, so that's one of the good strong points about it, that we know the area, we know the people who live in the area. It's alright for some people working here but at the end of the day they get in their cars and they go to their little nice housing estates, but we actually live on the estate. (interview, 22 March 2006)

Susan discussed the mutual support between people within the community, and then noted a tension between local people and outsiders who worked in the community centre. Indeed, it was precisely because of the strong community spirit and history of engagement with local politics that the regeneration process did not go ahead as planned in 2001 under the Newcastle City Council plan, Going for Growth.

Collective memory of a shared industrial past, in the form of community solidarity, is almost all that remains of a rich industrial heritage that grows dimmer with each passing generation. However, the notion of strong families and a strong community presents an idealized picture of life in Walker, and the tensions within the community are worth reflecting on in relation to this theme. While highlighting the strength of the community, local interviewees also referred to their fear of crime and vandalism, problems of drug and alcohol abuse, and rivalries between neighbourhoods (interviews with local residents, 12 September 2005; 22 March 2006). They lamented the decline of shops and services over the past thirty years and the lack of choice for young people in the area, and noted that available goods were overpriced and included little fresh fruit and vegetables. Recently, with the arrival of asylum seekers, notions of community solidarity and identity have been challenged by tensions between the traditional white working-class residents and the newcomers.

Asylum seekers are at the extreme margins of society, and so in some ways, there are grounds for solidarity in shared experiences (albeit on different scales) of deprivation. One of the founding members of the East End Community Development Alliance remarked that "a self-help culture in the community helps with the absorption of asylum seekers into the community" (interview, 1 December 2005). Nonetheless, he noted that there had been serious problems with racism, particularly in Shields Road and Church Walk, the two main shopping areas of Walker. Many such problems, according to several interviewees, went unreported. In the period since my field research, racial tensions have increased due to housing shortages and have led to a rise in votes for the far-right British National Party (BNP) in what has traditionally been a Labour ward (follow-up interview, Labour councillor for Walker, 8 July 2009). The tensions over community resources in Walker – between the pre-existing community and the new groups of primarily African asylum seekers – parallel some of the tensions discussed in the follow-up study of *The New East End* (Dench et al. 2006) between the "traditional" white, working-class community in Bethnal Green in the East End of London and the Bangladeshi community which has settled in the community since World War II.

A local Anglican Church minister took me to visit the dilapidated shopping area of Church Walk as part of a "church tour" of the East

End of Newcastle. It was a run-down square with just a few shops, and the minister told me that it was known for drugs, crime, and anti-social behaviour. The square was surrounded by tall, dingy apartment blocks, which were examples of the "hard-to-let housing" designated for asylum seekers. The minister showed me a charity in Church Walk set up by members of the community and voluntary sector, including the Anglican Church, to provide relief aid to asylum seekers in the form of clothing, household goods, and counselling and support services. Asylum seekers were the principle group to move into the area, which had a declining population, because they were placed there by the authorities rather than having their choice of a preferred locale.

I visited the charity again on one of the busiest days, Monday, since the premises were closed on the weekend. I met some of the staff and heard tales of the latest group of asylum seekers who had encountered difficulties with the government over staying in the country. I watched the staff help the asylum seekers fill out forms, write letters, and access online resources. I spoke with one asylum seeker from East Africa who said that he had been in Newcastle for two months but didn't like it. He said that his only social life in the city was at the charity. He cited incidents of racism throughout the community and city. His perception of the charity as the only friendly place in Walker highlighted tensions within the community between old and new populations.

In my discussion with two residents of Pottery Bank, I asked about the relationship between "local people" and asylum seekers. At first, they seemed reluctant to express their views. One resident began by saying, "Well there's loads. It's getting to be where it's like there are more asylum seekers than local people, but we work with them in here. We are doing cooking lessons with them here" (interview, 22 March 2006). The second resident interjected by commenting on the problem of local perception, whereby people thought that asylum seekers were unfairly given houses and microwaves. The two residents then laughed, saying "no comment" to waive their own views on "fairness," but then they elaborated on the ambivalent feelings within the community:

> I think it's getting better, people are getting used to them, and they're integrating more. We've got kids in the play group and in the school. A lot of the problem is like, that money is for the asylum seekers, and that money is for such and such. We're saying, "It shouldn't be like that. It should be there's the pot of money; it's for everybody."

On the whole, the two residents tried to give a balanced perspective. They showed sympathy for the asylum seekers, were willing to live and work with them, and explained tensions between local people and asylum seekers as a matter of competition over resources within the community:

> And a lot of it is ignorance, isn't it, because as I say, it's the same with them, there's good and bad amongst them as well. They're not all *dead canny* [North East English expression for a very kind and genuine person]. There's some that you think, God, they're a bit stuck up and things. And it's not our fault they've got to come here, but in the end of the day we've got to get on with it, we all have to live together. (interview, 22 March 2006)

This final comment encapsulates the spirit of community solidarity that so defines Walker. However, one could also interpret this comment as reflective of the more general attitude of resignation in the face of a long history of struggle with social, cultural, and economic difficulties within the community. Local residents relayed many stories about drug addiction, suicide, and people struggling to "get on with it" under different socio-economic situations. The accounts reveal tension between different social groups: between different areas and neighbourhoods, between working and non-working families, between men and women, between generations, and between the established white community and the new asylum seekers in the area. Yet there was also an attitude of "getting on with it" and "having to live together" which seemed to unite people in Walker, at least in the sense of everyday life and experience.

Community cohesion as living memory, as the present-day embodiment of a shared industrial past, is perhaps an overly romanticized concept, particularly in the context of a community faced with deep divisions and economic troubles. At the very least, it is possible to argue that there is unity amidst the decline, a solidarity that originates from a shared industrial past or a shared sense of dislocation in times of post-industrial change. There is an apparent contradiction between community spirit (solidarity and cohesion through strong families and networks) and community decline (as defined through socio-economic indicators). These two issues represent positive and negative sides of the socio-economic and cultural impacts of industrial decline: the positive side embodies a less tangible community spirit founded, at least initially, on a community with a strong industrial identity, and the negative side involves material degradation and socio-economic deprivation. The concept of community solidarity in Walker is not straightforward; it raises an interesting set of questions in relation to the construction of place, both by residents and by policy-makers and planners. Indeed, Amin's (2005, 614) criticism of the political use of the notion of "community cohesion" by New Labour in the United Kingdom as a means of redefining the social, whereby the idea of community and the local "has been re-imagined as the cause, consequence, and remedy of social and spatial inequality," is relevant in the context of this case study.

Conclusion

The period of my study in Walker captures the uncertainty of a prolonged period of transformation in which the death knell has long been sounding but the final blow has not been struck. This is a slow and painful process, and it is still underway. The "inner world of creative destruction" (Zukin 1991) in the various forms of memory in Walker, articulated with sadness, anger, humour, resignation, or transference, is an inner world which reflects many of the difficulties in shifting from a social space of industrial decline towards an undefined post-industrial dream. Walker represents but one example of a place caught between past and future, in which the present is imbued with and has not moved far from the past, and the future is at best uncertain.

There was no commemoration of the local industrial history of shipbuilding in the form of monuments, tourist shops, museums, or cultural activities based around old industrial sites in Walker, either along the riverside or within the surrounding community. At the time of my fieldwork, from June 2005 to March 2006 and before the closure of Swan Hunter in July 2006, there were no plans to create any. The spatial, political, and economic context of Walker Riverside was of physical separation from the community (gated entrances, few access points, and a main road acting as a barrier), and regulation in the uses of spaces. The politics and economics behind shipbuilding in the area as a whole have shaped the ways in which people relate to the sites and the scope for commemoration, museums, and cultural use. In this context, such processes could only occur in a top-down official capacity. The idea of a shipbuilding museum emerged only once the flagship yard of Swan Hunter had finally been ruined. Before, with 260 workers still employed, the myth that shipbuilding was alive in the North East could be sustained. It remains to be seen whether official memory in the future will mesh with local memory in Walker, and whether it can offer any kind of "closure" or gateway to a future, particularly one without jobs. After all, as Mark said, "we can't all sell each other baskets and jam." The regional government agency OneNorthEast is "backing a number of horses," including mostly "low value things like call centres," cultural industries, and tourism. But there is no single industrial sector that the hopes of the North East are being pinned on (interview, economic inclusion officer, 29 September 2005).

The forms of historical memory in Walker located within my interviewees' accounts were embedded in processes of social and economic change, and in how people understood themselves in relation to place and history. Shipbuilding and manufacturing in the North East have long been considered to be in irreversible decline, but socio-economic and psychological rupture with the past were held out in the maintenance of Swan Hunter as the last shipyard of the Tyne, and in the strength of community spirit shaped by industrial legacy which has remained etched in the socio-economic landscape of Walker. Boyer (1994, 133) reflects on Halbwachs' distinction between artificial (socially constructed) and "real" collective memory as follows: "As long as memory stays alive within a group's collective experience, he argued, there is no necessity to write it down or to fix it as the official story of events. But when distance appears, conferring its distinctions and exclusions, opening a gap between the enactments of the past and the recall of the present, then history begins to be artificially recreated." Following this logic, one might say that memory in Walker is only now entering a stage where it can be called upon and retold as a reconstructed past.

The narratives of local people in Walker reflected different forms of memory. There was a politics of memory that came through in the accounts of former shipyard workers, a politics that was anchored in the wider context of deindustrialization, Thatcherism, the decline of shipbuilding, and the declining voice of trade unions in the North East as a whole. These accounts were tinged with sadness, loss, and anger, but also with resignation and a sense that this story has become well-rehearsed. There were generational differences in memory, where stronger associations with the industrial past were often reflected in older generations. There was also a localized memory, which the narratives of local residents in Walker reflected, where memories could be both sad and funny, and where the industrial past was folded into the rhythms and fabric of the community. Many residents in Walker have resisted community regeneration rather than lamenting or protesting the abandonment of industry. In a way, they have protected their shared industrial memory through protecting their community. However, the hundreds of asylum seekers that have arrived in the area since 2000 have challenged residents' sense of solidarity and identity, and have produced new racial tensions, primarily based around competition over community resources. Finally, movements toward public commemoration of industrial heritage in the form of museums, cultural uses of old industrial sites, or the construction of monuments were absent at both the local and official levels during the period of my study. The ruins and the physical traces of the industrial era, including the homes that were constructed to house the shipyard workers, represent the beginning of the end of living memory, and their subsequent demolition or re-appropriation represents erasure or reconstitution. With the closure of Swan Hunter, and with an ambitious regeneration plan involving "major impact" for the community underway, it seemed in 2006 that prolonged processes of deindustrialization in Walker were finally coming to a close, opening the door for new possibilities.

However, for the residents of Walker, the prospects for a successful transition to a post-industrial future remain far from certain. There is an important post-script to this period of uncertainty between 2005 and 2006: the regeneration scheme in Walker has since collapsed in the context of recession, slow development, and government cuts (see chapter 7). By July 2009, progress was effectively at a standstill: demolitions had taken place but many of the new houses had not sold, leaving acute housing shortages in the community (several follow-up interviews, July 2009). Then, in March 2011, national funding for the Housing Market Renewal regeneration scheme was cut by the United Kingdom coalition government, six years earlier than expected. This sparked considerable concern about people in Walker and numerous other deprived communities who had been displaced or were living next to demolished or half-demolished houses, although blame was primarily placed on the economic recession rather than the overall regeneration strategy (Cooper 2011; Long and Wilson 2011). The failure of housing-led regeneration in Walker suggests that there are serious flaws in Angel of the North's vision of a post-industrial future, particularly in the post-2008 context of recession.

Chapter Four

"We Ruined Everything around Us, but We Couldn't Change Ourselves": Enduring Soviet and Textile Identities in Ivanovo

One of the most telling images of Ivanovo is a statue of Lenin standing next to a smokestack (figure 4.1). Lenin and the smokestack are situated in the centre of the city, not far from other semi-working, semi-abandoned textile factories and from other statues of Soviet leaders, including another one of Lenin, tall and magnanimous, just a few blocks away.

Ivanovo remains an old industrial city. At the time of my research in autumn 2006, the city had made few moves in the direction of the "post-industrial." Instead, Ivanovo was struggling to rebuild its crumbling textile industries and to reverse the process of industrial ruination. With the collapse of the Soviet Union, the integration of the Ivanovo textile industry into the global market economy resulted in very low foreign competitiveness and a dramatic decline in domestic demand (Kouznetsov 2004). The textile industry in Ivanovo came to a complete standstill during the early 1990s, which produced a striking landscape of deindustrialization as abandoned textile factories lined both sides of the River Uvod throughout the city and across the region. As a "mature" light industry with a lower rate of investment, technological innovation, and profitability than in other industries, the textile industry had little chance of long-term success, particularly as a dominant industrial activity (Morrison 2008, 65). Once the textile manufacturing centre of the Soviet Union, Ivanovo entered the global market economy with a high concentration of textile factories and very little economic diversification. However, there were few alternative prospects for industrial or post-industrial development in the depressed economic context of the 1990s. Despite having a narrow range of low-value products, very low profitability, and few chances for long-term survival, textile factories in Ivanovo adopted survival strategies aimed

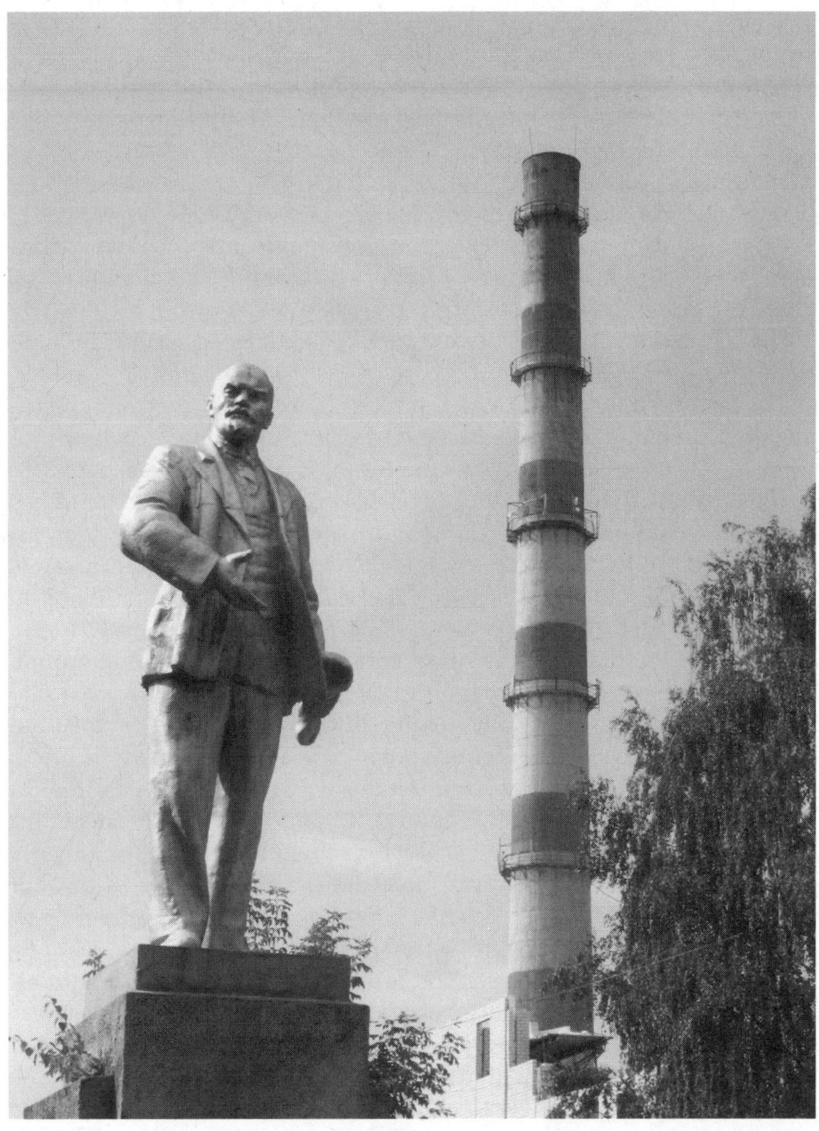

Figure 4.1 Lenin and smokestack, Ivanovo, September 2006

at facing immediate threats from foreign competition. These survival strategies were based on continuity with Soviet managerial practices, which were bureaucratic and concerned primarily with barter deals – relying on political ties and suppliers for financial support, in exchange for maintaining high employment and capacity – rather than market sales (Burawoy et al. 2000; Morrison 2008). Under this non-market-based logic of redevelopment, textile industries in Ivanovo gradually re-opened during the mid-1990s, albeit at significantly reduced capacities. There was a brief industry recovery after the 1998 Asian financial crisis, related to the collapse of the industry elsewhere, but by the early 2000s, the textile industry in Ivanovo once again faced serious difficulties (Morrison 2008).

The case of Ivanovo thus makes notions of market-driven deindustrialization problematic. There is a paradox in examining legacies of industrial ruination in Ivanovo, where the logic of the market has played a different role than in Western capitalist industries and the trend has been towards a partial reversal of industrial ruination rather than towards a post-industrial pathway. It makes sense to speak about legacies when something new has replaced the old, but it becomes difficult to separate the past from the present when the two are so closely fused. This echoes the theme of living memory in Walker, Newcastle upon Tyne, where the remains of shipbuilding lingered in the collective identity of the city. In Ivanovo, the textile industry was deeply embedded within the city's collective identity, not only as a source of historic pride, but also as the dominant economic engine for future growth, despite its failures in the global market economy. If the post-industrial ideal falls short of people's expectations in Western capitalist places like Niagara Falls and Newcastle upon Tyne, then how can this same ideal offer hope or possibilities for a place like Ivanovo, with such a different political, social, economic, and industrial history?

Ivanovo has maintained its textile and Soviet historical identities. This is evident not only at the everyday level but at the official level as well, for example, in *Ivanovskaya Oblast'* (Yefimov et al. 2006), a book advertising the Ivanovo Region that had just been published at the time of my research by the newly elected Ivanovo regional government. The first section in the book describes the city of Ivanovo, with the caption "Russian Manchester" set in bold letters (figure 4.2). This caption is accompanied by a central crest with an image of a female figure in traditional Russian peasant-style dress spinning cotton. The crest is flanked by two photographs depicting stone carvings of revolutionary figures from the Krasnaya Talka (Red Talka) Factory, located on the River Talka, northeast of the city centre. The factory was named "red" because it was the centre of the workers' struggles that led to the creation of the first Soviet of Workers' Deputies in Russia in 1905. Through its promotion of Ivanovo as the "Russian Manchester" and its inclusion of revolutionary figures from the Krasnaya Talka, *Ivanovskaya Oblast'* reveals that both Soviet and textile histories remain important within the present identity of the city.

Another tourist publication (Shushpanov and Oleksenko 2005) foregrounds a different popular city motto, "the City of Brides," and relates it to the romance of Ivanovo rather than to the historically high percentage of low-paid and unskilled female textile workers in the city: "After many years of trials Ivanovo got the most romantic name – 'The city of brides.' Ivanovo brides are not only beautiful, but also charming, well-educated and talented." This booklet, published by the Ivanovo city administration, guides the reader through different time periods in Ivanovo using a series of fictional love letters written from a young woman to a young man. The letters include selective details of historical and socio-economic changes, including positive "snapshot" representations of the Soviet era in 1917, 1934, 1942, and 1976. The final Soviet love letter reads: "1976. We are taking the lead over the Kamvolnyi Mill in the socialistic competition. Am now surpassing the work plan. The output program has been over-fulfilled by 200%. I've been awarded a trip for two around the Golden Ring by the local trade union committee." There is no account of post-Soviet conflict or change in the overall narrative. The final (and only post-Soviet) love letter is written on email using exaggeratedly modern slang, stressing technological rather than political or economic change: "Vovka, hiya! Read ur 2 emails ... Vovka u know I get crazy when I havent seen ya for 2 days ... See u at 5 by the lib. Love ya, kiss ya, cantwaittoseeya." This method of storytelling presents the history of Ivanovo as one of continuity and romance rather than of disruption and struggle, and it fails to mention the collapse of the Soviet Union.

Both publications emphasize the historical and present day importance of textiles in Ivanovo. They mention social and economic change over time, but they highlight socio-economic and cultural vibrancy throughout all periods and downplay the significance of post-Soviet transition. Despite pressures brought on by global capitalism for old industrial cities to move "beyond the ruins" (Cowie and Heathcott 2003) towards a post-industrial future, these publications demonstrate

ИВАНОВО
«русский Манчестер»

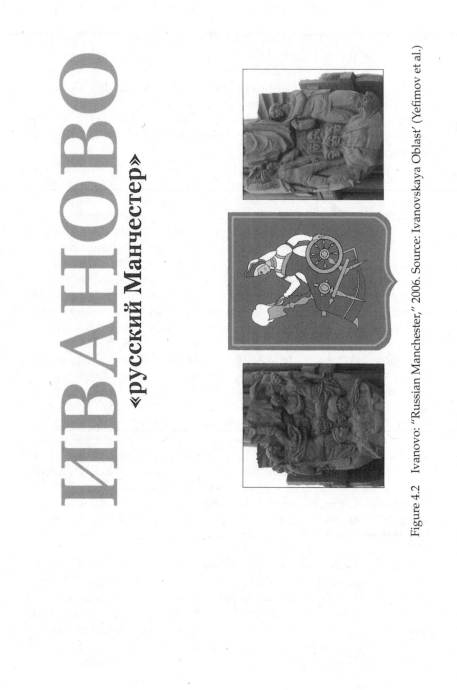

Figure 4.2 Ivanovo: "Russian Manchester," 2006. Source: Ivanovskaya Oblast' (Yefimov et al.)

that Ivanovo appears to be holding onto its image as a proud industrial city: the "Russian Manchester." This chapter will argue that, rather than ambivalent nostalgia over the loss of an industrial era of jobs and pollution as in the case of Niagara Falls, or sadness and uncertainty amidst imminent regeneration as in the case of Walker, living memory in Ivanovo was marked by both continuity and disruption with the Soviet and industrial past, evident in the tenacity of Soviet and textile identities despite significant ruptures to these identities.

Ivanovo: Setting the Context

Ivanovo is an out-of-the-way place. It does not fit into the nexus of global cities (Sassen 2002) or intermediate city hubs that define the landscape of global capitalism. The city is located just 300 kilometres northeast of Moscow, yet it falls firmly outside of the tourist map. For example, even though Ivanovo is one of the train stops along the Golden Ring of ancient cities around Moscow, it is excluded from the tourist draw of the area because of its identity as a polluted industrial city. In the year of my research in 2006, there was no entry for Ivanovo in the 2006 Lonely Planet guidebook to Russia. The "Way to Russia" online travel guide to the Golden Ring in 2006 even provided disclaimers about the tourist value of Ivanovo, such as, "in case you're stuck in Ivanovo and feel sad that the trip that was teaching you so much about architecture and history was abruptly paused in this town," and, "in case you like Ivanovo so much that you even decide to stay there, or (sorry) you're just stuck in the town, here's the list of a few hotels" (Way to Russia 2006).

Ivanovo was formed as an ideal socialist industrial city, and thus its story of industrial decline is linked to post-Soviet transition rather than to capital abandonment. The textile factories were established or taken over by the state during the Soviet period and developed as a concentrated textile hub within a centrally planned economy. There are numerous debates over the extent to which social and economic problems in the new Russia can be attributed to post-Soviet legacies or to the new capitalist context (cf. Ashwin and Clarke 2003; Burawoy and Verdery 1999; Chossudovsky 2003). The answer to this question is not clear-cut, as various factors point to both post-Soviet legacies and global capitalism. It can be said, however, that with the integration of Russia into a market economy and in keeping with the logic of capitalism, some industries have been winners and some have been losers, and textile industries in particular have been losers. During the Soviet period, a large number of textile factories in Ivanovo produced a narrow range of low-value cotton fabrics and light cloth for the state, a model of economic production that was impossible to sustain within the global market economy (Morrison 2008, 68). The role of capital in shaping the social and economic landscape of Ivanovo is most evident in the demise of industries, even if capital did not create or artificially sustain those industries. Indeed, the role of capital has become important in directing the future of industries in the region, both new and old. Since the end of the Soviet Union, there have been periods of complete industrial collapse, there has been a period of slow and partial recovery of old industry, and new industries have floated inconsistently in and out of the city.

The process of economic transition to capitalism in Russia has been widely theorized and debated (Burawoy and Verdery 1999; Freeland 2000; Kouznetsov 2004; Remington 2006). In particular, the unevenness of the transition has been described in much of the literature, with rise of the ultra-wealthy "oligarchs," the "victors of gladiator's capitalism" (Freeland 2000, 106), and the social and economic polarization of Russian society as a whole. In Moscow and St Petersburg, the juxtaposition between the very rich and the extremely poor has become stark, with flashy expensive cars, luxury restaurants, and designer shops (particularly in Moscow) in some parts of the cities, and high unemployment and decline in others. In Ivanovo, such contrasts are less evident, at least on the surface. There are some areas that are better off than others, but the city and region as a whole are marked by decline. This chapter follows the argument of Burawoy et al. (2000) that the post-Soviet transition can be characterized as a process of *involution*, to refer to market transition without the transformation of production, the society, or the state:

> Undoubtedly there has been a transition to a market economy but its consequence was not the *revolutionary* break-through anticipated by the prophets of neoliberalism, nor the *evolutionary* advance found in other countries such as China but an economic primitivization we call *involution*. On the one hand the market has impelled the degeneration of the manufacturing sector while on the other it has driven the vast majority of the population back on their own resources, intensifying household production. (Burawoy et al. 2000, 46)

Burawoy et al. (2000) highlight the survival strategies that people adopt for coping with the uncertainty of *involution*, for example through relying on kin networks, state support, and self-provisioning. Their analysis is relevant in the case of Ivanovo, where many people had to rely on produce from their gardens to survive during the mid-1990s; where factories have been kept afloat through barter exchanges; and where interpersonal and political connections have proved vital for economic survival. At the time of my research, the industrial (or post-industrial) future of Ivanovo seemed tenuous, hinged on old Soviet and industrial logics and practices, and compounded by long-term economic decline.

Industrial Ruination in Ivanovo

Industrial abandonment was abundant and pervasive throughout Ivanovo, rather than clustered in particular deprived areas as in the cases of Niagara Falls and Newcastle upon Tyne. The city centre of Ivanovo features two old textile factories, one on either side of the River Uvod. I recorded my first impressions of the southernmost factory (figure 4.3):

> From the river, I saw my first glimpse of industrial ruins – a massive red brick textile factory along the riverfront on the other side of the river, fenced in by walls, covered in graffiti, with some broken windows but generally structurally sound. A star was perched atop the tallest tower of the building. (3 September 2006)

As my field notes suggest, I thought that this factory was abandoned, but I later discovered that, like many other factories in Ivanovo, it was in fact partially in operation. On the other side of the Uvod stood the Big Ivanovo Factory, a flagship textile factory for the city, which faced serious economic difficulties and the possibility of imminent closure at the time of my research. Factories like these spanned the river in both directions for several miles; some were completely abandoned, but most were semi-abandoned. On the whole, it was difficult to tell from the outside whether factories were working or not. Although most factories appeared run-down, with broken windows and crumbling structures, they often maintained a minimally functioning area, and local people would not always be able to tell me with certainty whether a factory was fully abandoned or not.

Over the course of my research, I realized that the landscape of industrial ruination in Ivanovo was characterized at least as much by the phenomenon of partially working, partially abandoned factories as by completely abandoned factories. There are a number of factors behind

Figure 4.3 Semi-abandoned textile factory, Ivanovo, September 2006

this phenomenon. As previously noted, the number of textile factories has proven economically unviable in the context of the global market economy. However, the move towards the service economy or any kind of "post-industrial" economy has been slow, and few jobs have been created to compensate for the job losses in the textile and machine-building industries. Moreover, Ivanovo's textile identity is very strong. Nothing new has replaced the textile industries in any meaningful way, and as a result, many of the factories that closed in the 1990s began to re-open in a limited way as economic conditions slowly improved. Most of these factories were no longer profitable, and relied on Soviet-style barter exchanges (which, as previously mentioned, meant financial backing from political ties and suppliers in exchange for maintaining employment and capacity) rather than on market sales. Moreover, textile companies did not cooperate with one another: "Enterprises did not show any interest in or ability to develop a division of labour and persisted in manufacturing the same kind of goods, resulting in destructive competition" (Morrison 2008, 69). Partially working factories are part of a larger trend in post-Soviet Russia in which people continue to work, often without pay, in factories that have become insolvent and function at reduced capacities, and which in some cases are open for only a few months of the year (Kouznetsov 2004). Such factories continue to operate with a strongly gendered division of labour: a predominantly female, low-status workforce operating the looms and filling clerical positions, and "invariably" male managers, specialists, and heavy machinery workers (Morrison 2008). This reflects broader trends in the post-Soviet gendered division of labour in Russia (cf. Gerber and Mayorova 2006).

In textile factories in Ivanovo, production is organized into three lines: spinning, weaving, and finishing. I learned this from Andrei, a Russian instructor and former textile factory worker in Ivanovo, who offered to take me on factory tours in his Lada in lieu of Russian lessons (18 September 2006).[1] Andrei explained that a *kombinat* (industrial complex) is distinct from a *fabrica* (factory) because a *kombinat* is engaged in all three lines of production, whereas a *fabrica* is only engaged in one or two of the three lines. I was not given the precise definition of a *manufactura* (another word for factory) in relation to the number of lines of production. However, these distinctions have not held in the new economic reality: most of the *kombinati* in Ivanovo were engaged in only one of the lines of production. When I visited the partially working factories, I asked employees about the extent to which the factories were open. It became clear that there was mismatch between the extent to which factories were believed to be working (moderate and slowly improving) and the extent to which they were working in reality (barely at all and many were in serious crisis).

In the following sections I will focus on three old textile factories that illustrate different trends in the landscape of partial industrial ruination in Ivanovo: two partially working factories, Big Ivanovo Factory and New Ivanovo Factory, and one re-used factory, Silver City. I had great difficulty securing access to any of the partially working factories because of my status as a foreigner, a Westerner, and a woman, even through the use of gatekeepers such as my Russian instructor (and former textile worker), Andrei, and my student translators, Roman and Yuliya. My research was based on driving tours of the factories and on interviews with former workers and residents with connections to the textile industry.[2]

Big Ivanovo Factory

The Big Ivanovo Factory (Bolshaya Ivanovskaya Manufactura, BIM) is situated in the city centre, along the River Uvod and across from the central Pushkin Square. It is a large factory given its location – it virtually dominates the core of the city. Its buildings stretch back from the river; some are empty, while others have been converted into small businesses. The factory is one of the oldest in Ivanovo. Originally owned by a wealthy local entrepreneur named Burilin in the late nineteenth century, it was renamed the Company of Kuvaev's Factory after the Civil War (1917–23) and became the Big Ivanovo Factory in the 1970s. The factory is constructed of red brick, which was characteristic of industrial buildings in the late nineteenth century, with one tall smokestack dominating the skyline.

Like many other textile factories in the city, the Big Ivanovo Factory re-opened in the 1990s and has since struggled to survive. The local residents I interviewed were unsure as to the extent to which it was still working. On first glance, it appeared that the factory was not working at all, as many of its windows were broken and most of it seemed abandoned. However, a modern-looking Big Ivanovo Factory sign was posted on a lock over the river, and at certain times of the day smoke billowed out from parts of the rooftop. I was struck by the fact that the looming Big Ivanovo Factory and the semi-abandoned textile factory across the road were located in the centre of the city: these spaces seemed to be prime areas for downtown redevelopment. Perhaps related to this

Figure 4.4 Big Ivanovo Factory, Ivanovo, September 2006

fact, the factory had experienced numerous acts of sabotage and vandalism, and there was speculation that these acts had been done to encourage its closure. One resident, Elena, described these trends:

> This factory has more problems than other ones. It is in the centre of the city, and other people would like to get this territory to build something more useful. They say that it's a kind of sabotage. This factory has a lot of problems now. It's going to be closed up and resold and I've read in one of the papers that it's all done on purpose just to get the territory and to demolish the buildings and to build new ones for more profits. (interview, 26 September 2006)

I learned through follow-up correspondence with Elena in early 2008 that there had been a number of positive developments in Ivanovo since our first interview, including the reopening of the airport, increased investments, and the continued operation of the Big Ivanovo Factory, which seemed to be a local symbol for the textile industry in Ivanovo in the way that Swan Hunter was for the shipbuilding industry in Newcastle. These developments indicate that there may still be a future for textile industries in Ivanovo despite their many problems, or at least that the "survival strategies" of textile managers were still working in the short term.

New Ivanovo Factory

The comparatively small New Ivanovo Factory (Novaya Ivanovskaya Manufactura, NIM) is also situated in the centre of the city, just a short distance away from the Big Ivanovo Factory. The New Ivanovo Factory was formerly called the Zhidelyev Factory, after the local communist Nikolai Andreivich Zhidelyev. The local tradition during Soviet times was to name each factory after a communist revolutionary figure. The sign at the entrance to the factory shows evidence of post-Soviet renaming, as the old letters are still visible under the new ones. But the transformation was only half-hearted: at the time of my visit, Zhidelyev's plaque and ashes were still prominently displayed next to the factory entrance (figure 4.5).

I interviewed Tatiana, a former textile worker at the New Ivanovo Factory and the mother of Roman, one of my student translators (interview, 12 September 2006). During the Soviet era, Tatiana worked as a librarian in the New Ivanovo Factory, as factories in the Soviet Union had wider functions than industry. Each factory had its own "palace of

Figure 4.5 New Ivanovo Factory, Ivanovo, September 2006

culture" which hosted recreational and cultural activities and often housed a library to maintain its history. When the Soviet Union collapsed, one of the first things to go in the factories was this cultural side, so Tatiana lost her job as a librarian. She had to find an alternative way to earn money, which is the reason she became a textile worker when the New Ivanovo Factory re-opened. She described the work as hard and "personally destroying," both in terms of her own experience and in terms of her perception of the industry as a whole. Her negative impressions of the work were reinforced by her preferred occupation as a librarian rather than a textile worker.

At the New Ivanovo Factory, I was given special access to three members of the Veterans' Society, a group of retired female textile workers who held weekly meetings at the factory and offered social support to pensioners throughout the city. The Veterans were no longer employed at the factory, and so they were free to speak with me, which they agreed to do because they knew Tatiana and her son. The Veterans' Society met in a small room inside the factory that was filled with old pamphlets, books, and maps: the relics of the factory's former palace of culture. There was a large, bright-orange map of the Soviet Union on the wall that identified all the trade links between the cities which were associated with the textile industry. The three Veterans pointed out the extensiveness of the trade links and lamented the decline of trade connections with the end of the Soviet Union and the independence of the Central Asian republics. They gave a generally positive account of the history of the factory: they were proud of its role in the once massive textile industry in Ivanovo, and they were also deeply attached to the city itself despite its political, social, and economic problems. When I left, they gave me a book in Russian about the history of the factory, published in 1988, from their collection.

I found it interesting that the history of the factory in their room and their records ended at the end of the Soviet Union, despite the fact that the factory had re-opened after *perestroika* and continued to operate. The Veterans' room seemed to be frozen in time, and the material objects – the orange map on the wall, books published before 1988, and the construction of the shelves, tables, chairs, posters, and plans – all evoked a former era. The Veterans were only willing to share a united narrative; as a result, they spoke as a collective, with one designated leader (a former Communist Party spokesperson) and the others interjecting from time to time. They evoked spatial imagery in relation to their narrative of post-Soviet change. They referred to maps and objects in their room as coordinates of a particular worldview, and while many of their comments recognized social changes within the city, they also emphasized continuity with the industrial and Soviet past. The example of the New Ivanovo Factory offers insights into the cultural role of factories during the Soviet era, into the cultural as well as socioeconomic continuity between past and present, and into the close relationship between industrial and Soviet histories.

Silver City

There was one notable exception to the story of industrial ruination and decline in Ivanovo: the Silver City shopping centre. Silver City is a flagship conversion of the Eighth March Textile Factory into a shopping mall. Located in the city centre, Silver City represents the strongest physical evidence of the post-industrial in Ivanovo. Half of the factory's buildings were converted to a mall in the 1990s, and in autumn 2006 the second half was in the process of being converted into an extension of the mall. The Silver City shopping centre is very large, and includes four floors of shops and stalls, including a food court, a grocery store, and shops selling electronics, books, clothes, and many products one might find in any mall.

Treivish (2004, 17) writes that in Silver City, "one can feel here like elsewhere in the world." When I visited the shopping centre, I agreed that it certainly felt more similar to "elsewhere in the world" than it did to other places in Ivanovo. However, the mall did not appear to be post-industrial in an aesthetic sense. The outside of the building was covered with signs to mark it as a commercial space, but it still looked more like a textile factory than a four-story shopping mall, and no gloss had been applied to disguise this fact. Inside, the warehouse structure was still present, and many industrial elements remained exposed or unfinished. While many of the shops were typical of those in big box shopping centres, many others were arranged in the style of street market stalls. The aesthetic style of Silver City is fitting in the context of Ivanovo, which is more utilitarian and functional in its conversion than glass-fronted post-industrial examples. For example, it contrasts with wealthier urban areas where processes of regeneration and gentrification have transformed former industrial areas, such as Soho in New York City, the Guggenheim area in Bilbao, and Kreuzberg in Berlin. This suggests that there are multiple potential trajectories for post-industrial development, with multiple degrees of success. Indeed, since the time of my field research, Silver City has undergone subsequent

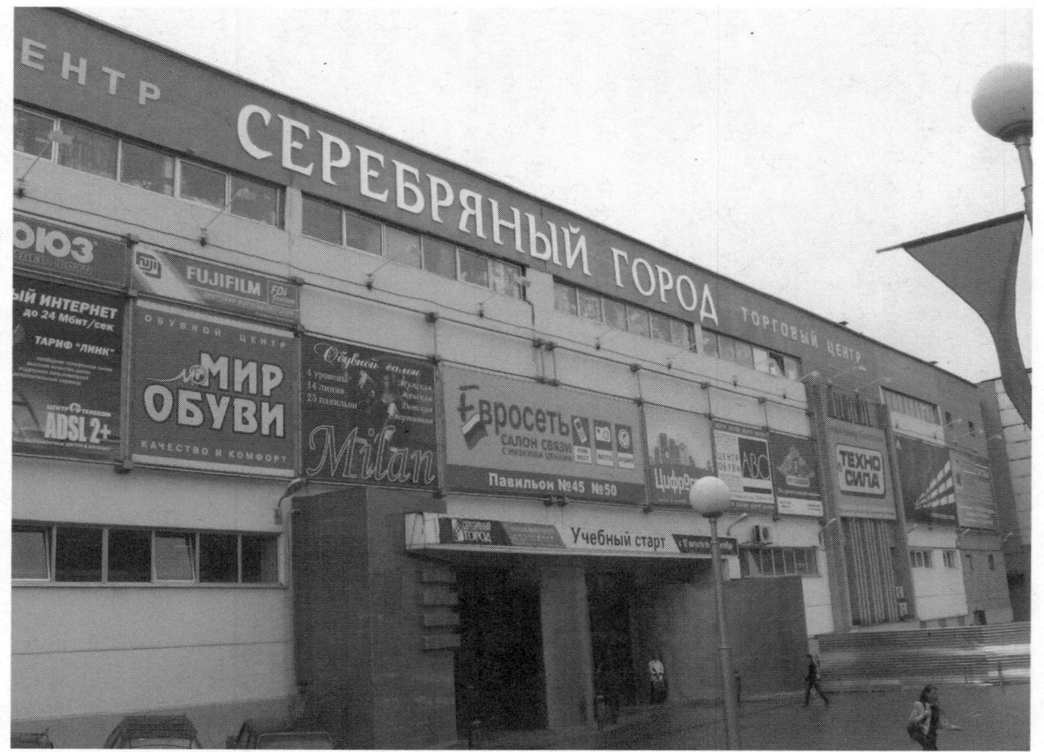

Figure 4.6 Silver City shopping mall, former Eighth March Textile Factory, Ivanovo, September 2006

redevelopment and the warehouse structure has been replaced with a more conventional post-industrial aesthetic.

Silver City is relatively unique in Ivanovo: such large-scale redevelopment of abandoned factories is the exception rather than the standard. Other examples of converted factories include textile shopping malls such as Textile Grad and Textile Profi, a night club in the city centre, and office spaces. Textile Grad is a former factory that has been converted into a textile market. Textile Profi is a utilitarian conversion of the Krasnaya Talka factory of the 1905 revolution. The factory has been divided into two parts: one half consists of the partially working remains of the original factory, and the other half is the textile mall. Such factory conversions represent a utilitarian use of space rather than a deliberate move towards post-industrial forms. While the shopping malls are places of consumption, they are firmly grounded in the historical textile identity of the city. They are organized more like informal street markets, with stalls and booths, than typical shopping malls. The other uses – as a night club, or as office spaces – more closely resemble the Western post-industrial prototype, but their economic viability remains uncertain.

In the midst of practical concerns about social and economic prospects in the city, many residents do not focus on the contemplation of derelict industrial sites as aesthetic objects or potential sites of culture. One resident, Tanya, described the visual impact of the textile factories of Ivanovo, one at the height of industrial strength and one at the point of ruination, as follows:

> It hurts me to see the Eighth March factory which was being built for a long time and very many efforts were spared on it. It was one of the unique factories, it was fully automated, and now it has been made into a shopping centre. There are enough shopping centres in our cities even without it. I feel the same sadness about Samoilovyi Kombinat [a textile factory with serious economic difficulties]. When I was a girl, I went there for an excursion. I was greatly impressed from seeing this building and all the production there. And now it hurts me to see that practically all the production there no longer exists. (interview, 21 September 2006)

In this statement, Tanya feels personally injured by the conversion of the Eighth March factory into a shopping mall and the near-abandonment of Samoilovyi Industrial Complex, another large factory in the city. Her husband, son, and extended family all worked in the textile industry at the time of our interview.

Form and process, the spatial, and the socio-economic blend and overlap in the landscape of ruination and partial ruination in Ivanovo. The textile industry was ruined in the early 1990s, and since then the process has slowly been reversed. Three themes emerged in relation to the particular character of industrial ruination in Ivanovo: (1) abundance and pervasiveness of industrial ruins throughout the cityscape, (2) form over function in the utilitarian use of space, and (3) the partial reversal of ruination, evident in the phenomenon of partially working factories and related to the tenacity of the city's textile industries. These themes connect with the legacies of industrial ruination and urban decline, which I will explore next, through illustrative narratives (cf. Sennett 1998) of local residents and former textile workers.

The Tenacity of Soviet and Textile Identities

Ivanovo has been slow to move away from identification with its Soviet and industrial pasts. Local people expressed their memories most clearly in relation to the Soviet past, and noted feelings of disjuncture at the separation between the old and new. This was expressed in the form of traumatic memory rather than as nostalgia, with a sense of dislocation and despair over throwing away the "good" along with the "bad," and a sense that the Soviet past still impacted mental attitudes. There were noticeable generational differences in the level of attachment to Soviet and industrial memories and identities. Local perceptions and experiences of industrial ruination in Ivanovo were deeply connected with the textile identity of the city, and while residents experienced sadness over decline, they also continued to use the old factories in pragmatic and utilitarian ways. Many people did not think of textile industries as belonging only to the past, to be remembered or commemorated, but rather saw them as part of the present. This suggests a form of living memory similar to that in the case of Walker in Newcastle upon Tyne, where the "battlefield" of deindustrialization had not yet healed over. The tenacity and resilience of the local population in the face of decline represents a powerful legacy of industrial ruination. This section will explore Soviet legacies and textile legacies in turn, and will focus on generational differences in accounts of Soviet continuities and disruptions, and on pragmatism and functionalism in relation to textile legacies.

Soviet Legacies and Generational Differences

Ivanovo was once an archetypal Soviet city, the only city to be placed alongside Leningrad (now St Petersburg) and Moscow as a "proletarian capital." However, unlike Ivanovo, St Petersburg and Moscow were historic cities of Tsarist Russia that possessed multiple cultural and socio-economic identities over the ages. When the Soviet Union collapsed, many historic symbols and spaces were reclaimed in both Moscow and St Petersburg, including Tsarist palaces and monuments, Orthodox churches, and public spaces such as Red Square in Moscow and the Winter Palace in St Petersburg. By contrast, the development of Ivanovo as a modern city occurred during Soviet times: before the twentieth century it was a quiet weaving village nestled in the forest. As a city with a thoroughly Soviet foundation, it proved difficult for Ivanovo to adapt to post-Soviet times. Many things in Russia were destroyed during the post-Soviet years, including the social, economic, and political infrastructure; factories; and ways of life. There were positive changes as well: the economy opened to the fluctuations of market capitalism, and the consumer products of a global economy became more available, as did greater access to media, travel, and casinos. The Russian anthropologist Oushakine (2007, 451) argues that "in the scholarship on cultural changes in post-socialist countries it has become a cliché to single out nostalgia as an increasingly prominent symbolic practice through which the legacy of the previous period makes itself visible." However, in Ivanovo there is no evidence of this form of typical Soviet nostalgia, otherwise exemplified in the *Ostalgie* (based on the German word *Ost* for "east") for daily consumer products from the former German Democratic Republic. Despite the end of the Soviet Union and the economic hardships that followed, Soviet legacies in Ivanovo remain pronounced, not only in the landscape of the city, but also in its social structure and in its people's ways of thinking.

Many of the residents I spoke with in Ivanovo described a "mentality" associated with the Soviet past, a psychological difficulty in coming to terms with the post-Soviet transition. Tatiana, the former librarian and textile worker from the New Ivanovo Factory, described the rupture between the present and the past caused by the collapse of the Soviet Union, and invoked the notion of ruins in the context of post-socialism rather than deindustrialization:

> All we had in the past was ruined and nothing new was built on the base of the old. When *perestroika* began, people tried to ruin everything that was connected with socialism, with the old life, but when they ruined everything they understood that not everything was so bad and some things could remain. It only became more difficult to build, to erect something new. We ruined everything and we couldn't build the new yet. Socialism was in our minds, not around us. We ruined everything around us, but we couldn't change ourselves. (interview, 12 September 2006)

This account highlights a connection between an interior landscape "in our minds," and an exterior world of *perestroika*. It also portrays this story as a collective struggle between interior and exterior forces. Tatiana expressed regret, for she believed that the "good" was thrown away with the "bad," and that aspects of the bad nonetheless remained. Her son Roman offered his view of this change, and pointed out that as a member of a younger generation that had seen only the results of the transition and not the process, he had not needed to adapt the way his mother's generation had to:

> It's one of the main problems of our parent's generation, so when they saw the two times, the older times and the new times, they lived in two different worlds. It's easy for me because I saw only the new world and I am used to it, but they had to be very flexible and some of them were a success and some of them were in a great depression. This depression can be seen even now among the older generation. (interview, 12 September 2006)

In this statement, Roman empathizes with the perspective of the older generation and admits that it has been easier for him than for his parents and grandparents. He refers to the tensions between the new and the old world, and the associated interior worlds ("a great depression") that come with such rupture. Roman belongs to the first generation in Russia that grew up in the post-Soviet period, but he was also influenced by the experiences of his parents and by traces of former Soviet life. As Oushakine (2000, 992) argues about this younger generation, "it is precisely this borderline socio-cultural location that allows us to see the generation of transition both as a product of current changes and as a symbolic manifestation of these changes."

Natasha, another resident and member of the older generation, expressed a similar view to Tatiana of what she termed a "Russian mentality" as a barrier to social change:

> The shift to new economic life was and is very difficult because it is very difficult to live in a new way and all the factors are combined together, and all the bad features from the past, from socialism, still influence us. And our mentality, the Russian, rather strange mentality, doesn't help us to shift. Moreover our people don't like anything new. (interview, 21 September 2006)

Echoing Tatiana and Roman's accounts, this description notes the tensions between old and new, and difficulty in erasing negative parts of the old in order to move forward.

I saw evidence of differences between generations in relation to the Soviet past in my encounters with people in Ivanovo. On a bus, one *babushka* (elderly woman) shouted at Roman and me for quietly speaking a foreign language: English instead of Russian. Other passengers defended us against her protest, and a lively debate ensued about whether we should be allowed to speak English or not. An old woman from the Veterans' Society at the New Ivanovo Factory expressed a similar sentiment about the use of foreign words in the city:

> I dislike a great expansion of some foreign words in the life of the city, and I don't like that many shops and other places are being named in a British or an American way and they often use English letters to write Russian words. They are Russian people. (interview, 21 September 2007)

By contrast, younger people appeared to be more open to Western influences, although very few spoke English. Most students at the university came from Ivanovo and nearby towns in Russia, or else from countries in Asia, Central Asia, and Africa. Apart from occasional European exchange students, there were few students from Western countries in the university. Perhaps for this reason, many local students expressed interest in me as a "Western" foreigner during my stay at the Ivanovo State University student residences, and they talked with me with about their dreams to travel outside of Russia. One local undergraduate student remarked that Ivanovo "can fairly be described as a city with no opportunities" (interview, 10 September 2006). She described her plans to work in the United States after her studies. Several students highlighted the high costs of travel abroad and the low wages and lack of local employment opportunities.

In Ivanovo, the Soviet past and the industrial past are fused. The break with both pasts has been sudden and disruptive, and yet social, cultural, and spatial aspects of each live on. There was a marked generational difference among my interviewees. The younger generation did not remember a time before the end of the Soviet Union, while the older generations did. Yet the younger generation carried a strong historical memory of the changes passed through the older generations and the relative "closeness" to the past. Oushakine describes the younger generation in Russia as "being caught in-between: between two classes (poor/rich), between two times (past/future), between two systems (Soviet/non-Soviet)" (Oushakine 2000, 995). In Ivanovo, the younger generation was closer to the former set of "in-between" categories, and this inherited historical memory of the Soviet past was underscored by the social and economic difficulties that young people have faced. In the older generation, there were strong feelings of loss related to the destruction of a former way of life combined with recognition that the past had contained both good and bad qualities. Finally, both generations expressed mixed feelings about what hope the future could bring.

Textile Legacies and Pragmatism

There was a certain pragmatism and functionalism in the ways in which people lived and worked within a city filled with ruins. Local identification with Ivanovo as the "Russian Manchester" was important for many respondents. Several in particular attached importance to the textile factories as the industrial heart of the city. Many factories continued to partially function, and nothing had replaced the loss of the textile industry, which might explain why the textile identity of the city remains so strong. Ivanovo has yet to find a path of economic development that can erase the past so deeply etched into its urban and social fabric.

Throughout the city, I noticed many references to textiles. There were many textile malls and informal textile markets throughout the city such as Textile Profi and Textile Grad, and there was also a museum devoted to the textile industry. The Textile Museum celebrated the history of textiles in Ivanovo, and at the time of my visit, it was divided into two sections, pre-revolutionary and post-revolutionary. However, like the books and maps at the New Ivanovo Factory, the museum's collection stopped abruptly at the end of the 1980s, and this was its only indication that there had been any disruption in the history of textile production in Ivanovo. Like the "City of Brides" booklet produced by the Ivanovo city administration, there was no marker for the post-Soviet period, nor was there any mention of the total collapse of the textile industry during *perestroika*.

The connection between work and industry sites in Ivanovo has remained integral to how local people think about the textile factories. Many interviewees, regardless of their age or degree of connection to the textile industry, mentioned that they associate the factories with hard work. One resident, an undergraduate student with no direct industry ties, had such an association even though she only related to the sites physically as a consumer, for instance at the Silver City shopping mall or the textile markets (interview, 20 September 2006). Another resident, a manager in the Textile Profi mall, also stressed that she associated the factories with hard work. She highlighted the importance of textile industries and blamed the government for the decline of industry:

> My near relations used to work in the Kohma textile industrial complex. The textile industry has been in decline for the last sixteen to twenty years. Now it's working, but the work has been reduced to a minimum. It's the government's fault that the factories are in decline; they did everything to bankrupt them. But these factories are very important for the development of the region. I associate these sites with hard work. (interview, 10 September 2006)

Tatiana, the former worker at the New Ivanovo Factory, also described her experience of working in the textile industry during the post-Soviet period in terms of hard work:

> Everything associated with my job in the textile factory was very hard work. It was like during pre-revolutionary times when the owner of the factory was the owner of the people and he made them do everything he wanted. I supposed myself to be a slave there. The work was very destroying, very hard. (interview, 12 September 2006)

For many people, the notion of hard work carried connotations of strength and resilience. But the association was not always a positive one, particularly for people who had experience with this type of work.

Several residents described problems in the city that they connected with industrial decline and socio-economic deprivation. They discussed the lack of employment opportunities, particularly for young people; high drug and alcohol use; inadequate health care, housing, and services; and poor roads and transport, among others. The physical evidence of these problems within the city and the personal accounts of hardship emphasized the depth and scale of the difficulties of everyday life in Ivanovo.

The connection between post-Soviet transition and industrial decline is similar to that between the environmental disaster at Love Canal and the onset of deindustrialization in Niagara Falls. For many people in Ivanovo, this was a traumatic experience, with devastating psychological and socio-economic impacts. For example, Natasha described her experience of change in Ivanovo as one that made her depressed:

> The 1990s were an extremely difficult time for all the people and for me as well. I was in a great depression and I didn't know how to go on living as it seemed to me like everything would be ruined and our country wouldn't exist anymore. I didn't know what to do, but now I suppose that the times are changing and our life is slowly becoming better. However, I must emphasize all the changes are very slow and these slow changes are influenced by the indifference of our people. (interview, local resident, 21 September 2005)

Natasha conveyed a sense of endurance and hope that the situation was slowly improving. However, she related the difficulty of adapting to social and economic change with general cultural indifference. This theme relates to the "Russian mentality," which Natasha described before, of a cultural block from the socialist past that makes accepting change more difficult. It also connects with the theme of weak political voice that emerged in accounts of the local politics of Ivanovo, which will be discussed in chapter 7.

Some people also spoke about their traumatic experiences in terms of social justice, and argued that their problems had been neglected by the state. For instance, Roman commented on the very serious problem of alcohol and drug use among young people in Ivanovo and its effect on his own situation as a young person:

> Alcohol is still one of the principal problems. It is born of the problem of a great deal of free time and the problem to do something in the evening ... Maybe the number of people who get some alcohol, it is practically the same, but the number of people who take drugs is increasing and the number of people who are infected with AIDS and other diseases connected with taking drugs is greatly increasing ... I suppose that the same problem leads to taking drugs, the problem that not so many things are done for the lives of the young generation. I can say that our government doesn't think of us, practically. (interview, 12 September 2006)

Roman attributed the high alcohol and drug use to the lack of social activities and government attention to the issues of young people in general. He described this as an injustice; in this sense, he seemed to speak not only for himself but on behalf of all young people.

Later, Roman and I visited Shuya, a small textile town that neighbours Ivanovo and is in the same region. Some of my interviewees described Shuya as an even better example of a depressed textile town. As the administrative centre of the region, with its own train station and a greater number of shops and services, Ivanovo has some activity apart from the textile industry. Shuya, however, does not, and its industry is in decline. For this reason, Roman and I took a *marshrutka* (routing taxi) one Saturday to visit the town. Not surprisingly, we did not get further than the gates of any of the factories in the town, and we were questioned by the security personnel when I took a photograph of one of the factories. After we walked the full length of the town and enquired at the main textile factories, which took just over an hour, we abandoned our mission for the day. Instead, we drank tea and ate some packaged biscuits in the only open local café. Looking at the disco ball attached to the ceiling, Roman remarked that he sincerely hoped that this establishment was not the only place where young people could go to have fun.

Natasha and Roman's accounts are compelling not simply because they are stories of hardship and trauma, but because they are also stories of resilience and modest hope for the future. Together with the accounts which emphasize hard work in relation to the textile factories, these accounts suggest that people in Ivanovo engage with legacies of industrial ruination with tenacity and endurance.

Conclusion

Industrial ruins are part of daily life in Ivanovo, and are suggestive of social and economic depression, decline, and blight. At the time of my fieldwork, there had been no move towards converting the space into an art gallery or museum, most simply because the money did not exist for such projects, the state had not promoted regeneration, and the landscape of industrial ruination was vast. Even the Silver City complex was utilitarian in its conversion. There had been no attempt to re-design the interior or exterior of the building, which both still resembled an old factory, and the signs for shops were arranged in an anarchic pattern. Similarly, I found no suggestion of a tradition of an artist, musician, or activist squatter community in Ivanovo, as in abandoned industrial sites in other cities around the world (cf. Florida 2005; Landry and Bianchini 1995; Laz and Laz 2001; Zukin 1991). Since the landscape of ruination in Ivanovo was vast and unregulated, many informal uses – such as drug and alcohol consumption, gathering, graffiti, and vandalism – were possible, but were not within the scope of my research. The only reference to such uses was the account of vandalism and sabotage already mentioned in the case of the Big Ivanovo Factory. I didn't witness any squatting or illegal activities, just consumption of alcohol in and around the ruins, some graffiti, and possible signs of occupation. For example, at the first semi-abandoned textile factory that I encountered in the city (figure 4.3), I noticed people strolling on either side of the river near the abandoned factory, a mixture of families and people of different ages and backgrounds. There were broken alcohol bottles on the ground and some people were drinking while walking (field notes, 3 September 2006). The only evidence of a link between art and ruined factories was found in a wall painting on the textile factory, which depicts an industrial landscape of empty rail lines (figure 4.7).

Perhaps, over time, Ivanovo will capitalize on its unique landscape of industrial ruination as a cultural and aesthetic commodity, as have other old industrial cities, such as Łódź, the "Polish Manchester" (Kaczmarek and Young 1999). However, aesthetic and cultural approaches towards industrial ruins, either in the realm of arts and journalism, or in the realm of economic development, tend to obscure the harsh realities of industrial decline. They romanticize these sites and, through that process, turn them into commodities: a form of creative destruction. Even after a few days of staying in Ivanovo, the embedded socio-economic difficulties eclipsed my awe at the scale of ruination. Experiences of industrial ruination in Ivanovo provide insights about how people get by with daily living in difficult material conditions. This theme relates to Michel de Certeau's (1984) analysis of "tactics and strategies" for daily living as a way of subverting dominant standards and expectations in the practice of everyday life. Socio-economic deprivation and industrial ruination define the landscape of Ivanovo as a whole, as a physical and as a lived space, and as a "real-and-imagined" (Soja 1996) place.

Paradoxically, changes in Ivanovo have occurred both too quickly and too slowly: the Soviet and industrial foundations of the city were erased overnight, and yet they persist in the city's identity and fabric. The contradictions between old visions and new realities are related to the city's unique past. The city has been transformed from an ideal

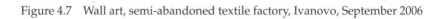

Figure 4.7 Wall art, semi-abandoned textile factory, Ivanovo, September 2006

socialist industrial city to an exemplar of industrial decline. Nothing has effectively replaced the positive socialist and industrial associations of the past, so many of these associations have survived. This part of the story of industrial ruination in Ivanovo is one fragment in a longer story with many possible trajectories and outcomes. In many ways, it parallels the early stories of industrial decline twenty to thirty years ago in places such as the North of England. Whether Ivanovo will or can follow the post-industrial paths of regeneration and the service economy like many other old industrial cities, whether it will remain an area marked by industrial decline, or whether it will find other alternatives, remains unclear.[3] Uncertainty defines the lives of people living in places situated precariously between moments of destruction and moments of recreation, including all three case studies in this book.

PART TWO

Themes of Industrial Ruination

Chapter Five

Reading Landscapes of Ruination, Deprivation, and Decline

Landscapes of industrial ruination constantly shift through cycles of decay, reuse, demolition, and redevelopment. My study of these landscapes captures brief slices of space and time within much wider processes of creation and destruction. One of the aims of this book is to unpack what particular spatial-temporal moments of industrial ruination can reveal of wider social and economic processes. Such an analysis is difficult because it involves reading processes within processes, rather than processes within fixed forms. The spatial-temporal moment, the "slice" or "snapshot" of ruination in each case, becomes an important focal point: the moment captures the process of ruination in time and space, making a finite sequence of ruins visible. In each of the cases under study, landscapes of urban decline and deprivation are integrally linked to landscapes of industrial ruination, and this chapter seeks to explore these interconnections. By interpreting landscapes as both form and process, spatial and social, cultural and material, this research aims to break down binaries of urban investigation. My methodological approach for reading landscapes of industrial ruination draws on the idea that social and spatial phenomena are interlinked, for the spatial is socially constituted and the social is also spatial (cf. Lefebvre 1991; Massey 1995; Soja 2000). However, conducting social and spatial analysis also presents specific challenges, since they are complex and occur across not only time and social relations but also space. As H.C. Darby (1962, 2) argues, "a series of geographical facts is much more difficult to present than a sequence of historical facts. Events follow one another in time in an inherently dramatic fashion that makes juxtaposition in time easier to convey through the written word than juxtaposition in space." Massey (1994) elaborates on the problem of "geographical description," for "in space you can go off in any direction and ... in space things which are next to one another are not necessarily connected" (267).

In recent years, there has been growing interest in visual methods in social and cultural research, and a number of texts have emerged on forms of visual analysis (Banks 2001; Emmison and Smith 2000; Knowles and Sweetman 2004; Pink 2001; Rose 2007). These texts are primarily concerned with analysing pictorial representations, including photography, film, video, and maps, often produced by research participants and interpreted as ways that people understand their social environments. Knowles and Sweetman (2004, 7) provide a relevant framework for thinking about visual methods that is inspired by C. Wright Mills' (1959) idea of the "sociological imagination":

> In capturing the particularity of social processes, illustrating the general and the particular, and illuminating the relationship between the two, visual methods are particularly well suited to developing what Mills referred to – in what was intended in a non-disciplinary-specific sense – as the "sociological imagination." The sociological imagination links the larger historical and social scenes in which lives are set with individual experience and biography.

In the spirit of the sociological imagination, visual methods are well-suited to analysing industrial ruination as a lived process. While this book analyses landscapes rather than pictorial representations, many of the aims and methods are similar – for example, interpreting visual objects rather than texts, and analysing other people's understandings of these objects. There are also some key differences, as my work focuses on spatial analysis of objects in three-dimensional space, rather than two-dimensional representations. There has been relatively little research on how to undertake spatial analysis, although a great deal has been written more generally on "space" and the city (Lefebvre 1991; Massey 1994; Soja 1996).

My approach to spatial analysis was intuitive and investigative, in some ways similar to the methods of "psychogeography" exemplified in the popular work of London author Ian Sinclair (cf. Sinclair 2002), who walked around the motorways and neighbourhoods of London to trace connections between history, psychology, and urban geography. My work also relates to an emerging form of social and spatial analysis that involves "mobile methods" that capture, track, simulate, and "go along with" moving experiences and systems in the contemporary world (cf. Anderson 2004; Büscher and Urry 2009; Urry 2007). As Anderson (2004, 254) argues, "'talking whilst walking' can harness place as an active trigger to prompt knowledge recollection and production." Indeed, talking while walking and, by extension, while driving, while on the train, or while using other transportation, are useful for researching the relationships between people and places. My mobile methods included tours of old industrial areas while walking, driving, or on public transport, both alone and with research informants.

The theoretical framework of "industrial ruination as lived process" addresses the relationship between people, places, and historical and socio-economic processes. In addition to visual, spatial, and mobile methods, ethnography is particularly useful for studying historical and socio-economic processes, for example in cases involving transition or change, given the fleeting nature of different moments in space and time. Burawoy and Verdery (1999, 2–3) describe the special relevance of ethnographic approaches for understanding processes of transformation, using the example of post-socialist transition:

> It is precisely the sudden importance of the micro processes lodged in moments of transformation that privileges an ethnographic approach. Aggregate statistics and compendia of decrees and laws tell us little without complementary close descriptions of how people – ranging from farmers to factory workers, from traders to bureaucrats, from managers to welfare clients – are responding to the uncertainties they face. From their calculations, improvisations and decisions will emerge the elements of new structurings. Thus, even an ephemeral moment captured ethnographically will reveal something of the conflicts and alternatives thrown up by the destructuring effects of the end of state socialism.

This book draws on mixed qualitative methods combining social and spatial analysis to capture such ephemeral moments, and people's different responses to uncertainties they face within these moments was one of the key themes which emerged from the research.

Through visual, spatial, and mobile methods, including site observations, walking and driving tours, and photography, as well as in-depth narrative interviews and ethnographic observations, this chapter aims to read spatial, socio-economic, and temporal dimensions of industrial ruination. The analytical task of reading sites and processes of industrial ruination presented me with the challenge, often found in ethnographic work, of getting close to sites and processes of industrial ruination, yet avoiding the outsider perspective of a dereliction tourist or a voyeur. The difficult balance between perspectives of distance and proximity has been widely theorized in sociological writings on the city. Lefebvre (2003, 117) writes of proximity as central to creation and production in the city: "Nothing exists without exchange, without union, without proximity, that is, without relationships." He associates distance with tension and violence in the city: "(Social) relationships continue to deteriorate based on the distance, time, and space that separate institutions and groups. They are revealed in the (virtual) negation of that distance. This is the source of the latest violence inherent in the urban" (2003, 118). Similarly, Michel de Certeau (1998) criticizes a top-down bird's-eye view of the city, and instead advocates a way of experiencing and being in the city that embraces the commonplace and the everyday. Both Lefebvre and de Certeau privilege proximity over distance as perspectives on the city. By contrast, Benjamin's "flaneur" (2000), a term borrowed from Baudelaire, is a detached urban observer who strolls through the streets of a city, keenly perceptive yet uninvolved, close yet far away. Benjamin's flaneur engages with aspects of both distance and proximity through embracing notions of detachment and propinquity. Simmel (1997) also writes about the effect of spatial conditions on forms of social, psychological, and physical distance. He argues that increases in physical proximity between individuals, particularly through concentration of people in cities, can lead to overstimulation. As a reaction to being too close to others, people in cities learn to adopt a "blasé metropolitan attitude" that is detached and socially distant.

Distance and proximity represent two sides of the same picture, and during my research I did not privilege one over the other, but rather used them both as a guide for navigating and understanding some of the complex dimensions of industrial ruination as a lived experience. This was particularly important given that my research methodology relied a great deal on intuition: following various pathways and roads across cities; poking around abandoned factories which were contaminated, fenced off, or guarded; and finding informants to accompany me on tours of former industries and devastated communities. My aim was to read these landscapes of industrial ruination and unravel their multiple meanings like palimpsests.

A palimpsest refers to a manuscript in which old writing has been rubbed out to make room for new writing, or a monumental brass turned over for a new inscription. The metaphor of place as a palimpsest is often invoked as a theme within urban sociology, human geography, and contemporary archaeology (Buchli and Lucas 2001; de Certeau 1984; Harvey 2003; Van der Hoorn 2003). As Michel de Certeau (1984, 201–2) writes:

> Beneath the fabricating and universal writing of technology, opaque and stubborn places remain. The revolutions of history, economic mutations, demographic mixtures lie in layers within it, and remain there, hidden in customs, rites, and spatial practices. The legible discourses that formerly articulated them have disappeared, or left only fragments in language. This place, on the surface, seems to be a collage. In reality, in its depth it is ubiquitous. A piling up of heterogeneous places ... The place is a palimpsest.

In other words, "place is a palimpsest" because of relationships between forms and processes, between the spatial and the temporal, and between social, historical, economic, cultural, and urban dynamics. Places are layered with traces of celebrations and disasters; with different patterns of streets, buildings, and trees; and with memories. In the edited volume *Archaeologies of the Contemporary Past*, Buchli and Lucas (2001) highlight the value of the anthropological study of modern-day archaeological deposits, which are generally seen as waste. The essays in their book explore a range of archaeological sites and uncover patterns of consumerism, political and social change, collective memory and forgetfulness, and residue in places impacted by war or economic abandonment. While many urban scholars have described places as palimpsests, few have set out to unravel them, to unpack their meanings in the manner of contemporary archaeology. This chapter takes up this challenge through the process of reading industrial ruination, manifested in social, spatial, and temporal landscapes.

David Harvey (2003, 230) reflects critically on the implications of understanding the city as a palimpsest, and asks questions about how the different forms, processes, meanings, and possibilities are interconnected:

> The result [of "the relationship between the urbanizing process and this thing called the city"] is an urban environment constituted as a palimpsest, a series of layers constituted and constructed at different historical moments all superimposed upon each other. The question then becomes how does the life process work in and around all of those things which have been constituted at different historical periods? How are new meanings given to them? How are new possibilities constructed?

Harvey suggests that the idea of the city as a palimpsest raises important questions, not only about historical and material processes, but also about the relationship between the social "life process" and urban palimpsests written into the fabric of the city. In the spirit of Harvey's inquiry, this chapter explores the interconnections between landscapes of ruination and their adjacent communities and cityscapes. It aims to break down distinctions of form and process, social and spatial, spatial and temporal, landscapes and legacies: to read between and through the layers. By identifying a key theme that links landscapes of industrial ruination to landscapes of community – the spatiality of deprivation and social exclusion – this chapter also provides context for considering further cross-cutting themes across the three cases.

Reading Landscapes of Industrial Ruination

Can industrial ruins tell their own stories? Is it possible to interpret the flows of capital, people, and places by looking at an abandoned shipyard, an abandoned textile factory, or a vast empty field where a chemical factory once stood? Does it matter if that shipyard is in Newcastle, in Glasgow, or in a city along the Indian Ocean? Does it matter if the abandoned factory produced textiles or chemicals? Does it matter if the ruins are visible or unseen? Some traces and clues are indeed visible, but one needs to know where and how to look for them. Tim Edensor (2005) implores us to think about industrial ruins without focusing on their particular geographical locations, to experience their universal qualities. My research was concerned with the opposite: drawing connections between landscapes of industrial ruination and their specific contexts and geographies. One of my aims was to examine industrial ruins as artefacts of social and economic processes rather than as aesthetic or cultural objects. Could industrial ruins be read as the waste products of capital abandonment or capital integration into new markets; as the footprint of uneven capitalist development?

The chemical brownfields of Niagara Falls, the vacant shipyards of Newcastle upon Tyne, and the partially abandoned textile factories of Ivanovo offered different insights into reading landscapes of industrial ruination. On one level, it was difficult to read anything in the physical landscapes themselves. It was rarely clear whether a shipyard had closed due to capital moving elsewhere, capital insolvency, or a lack of state contracts and subsidies. The textile factories that had been abandoned as a result of post-socialist transition to a market economy looked much the same as chemical factories that had been abandoned by capital flight. Derelict chemical factories with high levels of suspected contamination looked no more dangerous than those without.

Nevertheless, there were markers and clues written in the landscapes. The fenced-off overgrown field at Love Canal marked with a sign warning "do not enter" and surrounded by roads leading nowhere gave off an eerie feeling – what had happened to this place? The fenced-off field of the former Cyanamid site, where the only building left was a boarded-up, graffiti-tagged community centre, gave off a similar sense of neglect. Other clues within the landscape of the twin cities of Niagara Falls were the derelict downtowns, the socially excluded and impoverished residential neighbourhoods, and the large number of abandoned industrial sites scattered across the region. In the Highland Avenue community, the spatial clues of devastation were particularly striking: the looming derelict factories with broken windows and tall weeds cast shadows over the community. The streets were full of potholes, the stores were run-down, the factories were visible from all angles within the community, and yet the houses and churches were relatively well-kept. These houses and churches suggested that despite the devastation, people cared about their homes and their community. In Walker, the strip of industrial riverside dotted with offshore companies, cranes, warehouses, scrap yards, and empty shipyards indicated dereliction and decline. However, the fences around the industrial riverside area and the continued existence of mixed industries suggested that industrial life on the River Tyne had been prolonged and regulated. The road separating the industrial riverside from the red brick council houses in the residential community suggested a disconnection from the present-day industrial riverside and the adjacent community. Finally, in Ivanovo, the sheer abundance of textile factories, their tightly guarded gates, the Soviet murals on their walls, the ashes of former communists, and the use of former factories as textile markets all pointed to the particularity of the post-socialist context. The abundance of textile factories also suggested that there were far more mono-industrial factories clustered in a single city than might have been developed even in a Dickensian capitalist setting, which hinted at a legacy of central planning. Visual clues such as billowing smoke, security personnel, and vehicles in parking lots were evidence that many textile factories in Ivanovo were starting to function again, despite the broken windows and dilapidated infrastructure.

Through reading the visual clues within vacant shipyards, abandoned factories, and brownfield sites, a wider picture began to emerge, connected to stories of deindustrialization and uneven capitalist development, but also to stories of regulation, environmental disaster, demolition, and reuse. My research process began with reading physical landscapes and slowly developed to capture broader social, economic, and historical landscapes through interviews, documentary research, and ethnographic observations. These initial readings were important guidelines for a process of further discovery and unravelling the multiple layers. Landscapes of industrial ruination are also deeply interconnected with the social life of the residents and workers in cities and communities around them. Drawing on a combination of spatial and ethnographic methods, the next section will investigate "how the life process work(s) in and around" (Harvey 2003) industrial ruination in each of the three cases.

The Spatiality of Deprivation and Social Exclusion

Landscapes of industrial ruination cannot be considered in isolation from their wider social contexts and environments. In reading the social and spatial landscapes, I examined the interconnections between landscapes of ruination and landscapes of urban community. Two related themes emerged across all three cases: (1) socio-economic continuities between industrial ruination, urban decline, and community (or urban) deprivation, and (2) uneven geographies of socio-economic deprivation, with contrasts between "rich" and "poor" areas evident at different scales (regions, cities, and communities). Taken together, these form one overarching theme: the spatiality of deprivation and social exclusion. As Byrne argues (1999, 1), social exclusion is inherently dynamic, for "exclusion happens in time, in a time of history, and 'determines' the lives of the individuals and collectivities who are excluded and of the individual and collectivities who are not." Social exclusion is also inherently systematic because "exclusion is something that is done by some people to other people" (Byrne 1999, 1). Madanipour (1998) argues that social exclusion often has clear spatial manifestations in deprived inner city or peripheral areas. This "spatiality of social exclusion" (Madanipour 1998) relates to the concept of a classic ghetto, "the involuntary segregation of a group which stands in a subordinate political and social relationship to its surrounding society" (Marcuse 1997, 228). There are many forms of social exclusion that are not spatially concentrated, such as homelessness, or areas in which social exclusion has been intentionally dispersed through slum clearance programmes and other forms of housing management (Madanipour 1998). The theme of the spatiality of deprivation and social exclusion, which links landscapes of industrial ruination with landscapes of community, will be addressed in greater detail in relation to each of the cases.

Niagara Falls

A significant theme that emerged in my analysis of the relationship between old industrial sites and adjacent communities in Niagara Falls was that of discriminatory or exclusionary spatial segregation on the basis of race and class, by pushing certain social groups to live in areas with increased health and safety risks. Both Highland and Glenview-Silvertown are clear pictures of spatialized social exclusion, not only through the clustering of socio-economic deprivation, but also through residential exposure to toxic pollution. The Highland and Cyanamid brownfields are both adjacent to poor, working class areas. The Highland residential community is also predominantly African American. The old industrial sites in both cases are so close to residential areas that they are part of them: Cyanamid is part of Glenview-Silvertown, and the Highland Avenue brownfields are located along the main artery of the Highland community. The old industries are woven into – and have socio-economic continuities with – the fabric of the surrounding neighbourhoods. The landscapes of ruination and their adjacent communities in Niagara Falls stand in juxtaposition with wealthy, white areas on the New York side and with tourist areas on the Canadian side, reflecting an uneven geography of capitalist development. Class and race were major themes in the spatial and social exclusion of people in these areas.

On my many journeys to the Highland community, I noticed clear spatial indicators of the separation of Highland from the rest of Niagara Falls: one must cross railroad tracks over the two main roads which lead into the community, which itself is located at the boundary of the city. The environmental and racial segregation in Highland arguably represents an example of "environmental racism," which in general terms refers to racial discrimination whereby polluting industries, toxic dumps, or other environmental hazards normally deemed unsuitable for residential proximity are located next to minority groups (Bullard 1993; Clairmont and Magill 1974; Cole and Foster 2001). One historical example of "environmental racism" is the community of Africville, an Afro-Canadian neighbourhood in Halifax that was first settled in the 1700s and grew to be surrounded by toxic polluting industries. In the

1960s, the city forced residents to relocate for a small sum of money as part of an urban renewal project, a process which many criticized as "ethnic cleansing" (Clairmont and Magill 1974). The African American community that grew up around the factories of Highland Avenue, and was prevented from living in other parts of the city, could be described as subject to a form of environmental racism.

Several residents in Highland Avenue expressed awareness of their spatial and social exclusion as the result of racism. For example, Dan, a middle-aged African American man who had lived in Highland for his entire life (introduced in chapter 2), argued that the city's placement of an African American community in the area was a deliberate act of segregation: "The community is not accidentally segregated. The whole intention of the building of that community was a segregated community. These are railroad tracks, which are traditionally boundaries to separate communities" (interview, 21 March 2007). Dan said that in 1943, the city council made a decision to build public housing to address the housing shortage in the city:

> So now, those were poor people and largely, poor African Americans. Where do you put them? Here [pointing to a map]. Centre Court [public housing project] was supposed to be on 14th and Buffalo Avenue, down the street here. It was supposed to be near a church called Holy Trinity where they still conduct mass in Polish today. The Poles looked at the fact that they were going to put up a housing project that was largely for African Americans, next to their church. They protested vehemently. And so what happened is, the city decided, "Well, we'll put this someplace where we'll get the fewest amount of complaints." This was all a swamp. So they developed this neighbourhood and put this housing project here because you have the least amount of protestation.

Dan's reference to a particular image – one of the many maps of Highland in the corner of the Niagara Falls Public Library, where our interview took place – connected image and narrative. He defined in spatial terms the social exclusion which he felt that the Highland community had experienced, and he could point to these features on a map. We could have a spatial discussion, in this sense, just as I did while accompanying informants on driving and walking tours of other sites.

A former Highland resident, Carl, took me on a driving tour of Highland Avenue and pointed out the markers of separation within the community, including the railroad tracks and the city boundary to the

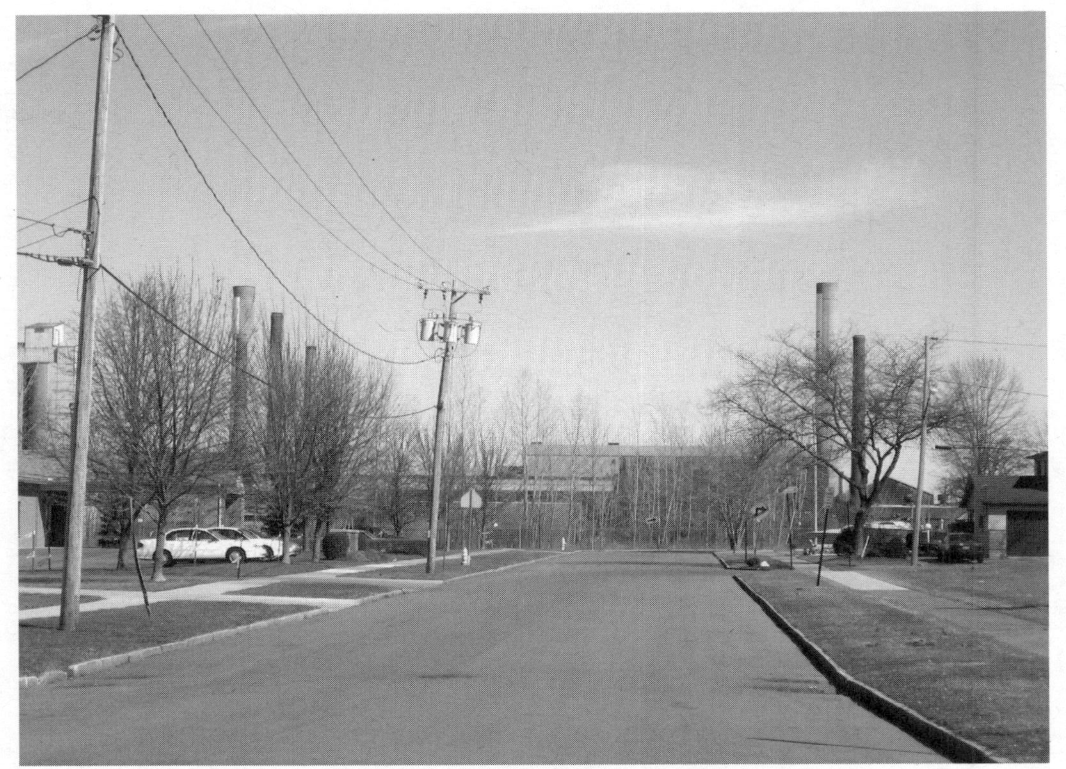

Figure 5.1 Smokestack views, Highland Avenue, Niagara Falls, March 2007

northeast (interview, 23 March 2007, also introduced in chapter 2). We drove over the tracks to explore adjacent neighbourhoods that gradually became wealthier and more white as they moved away from the factories and up along the Niagara Gorge. Finally, Carl drove to the neighbouring city of Lewiston to show me the contrast with a much richer area, including mansions overlooking the Gorge and a private golf course. He said that the only black family to have moved there had had their house fire-bombed a few years previously.

While many residents acknowledged the local de facto racial segregation, the notion of Highland as the product of environmental racism was something which seemed to belong more to the past, to the telling of the history of the area. Despite a general recognition of historical and present-day racial injustices, no residents or activists seemed concerned with making formal or legal claims of environmental racism.[1] In part, this may be because there are other cases of social and spatial marginalization in the city which do not correspond to race. Directly adjacent to Highland Avenue, also on the edge of the city and in close proximity to contaminated sites, there is a white community that has similar socio-economic problems of poverty, poor health, and stigmatization. The existence of that community suggests that spatial discrimination in Niagara Falls was based on not only race, but also class. However, the lack of political or legal action on the issue of environmental racism in Highland relates primarily to the fact that there is no one who can be readily held accountable for the contamination of the area. The corporations left long ago, and Niagara Falls City Council is a fraction of its former size, has a limited tax base, and is struggling to address depopulation, socio-economic deprivation, and crumbling infrastructure across the city.

Environmental racism did not seem to factor significantly into the Glenview-Silvertown case, as the area was predominantly white during the time of Cyanamid's operations.[2] Arguably, the Glenview-Silvertown community, which formed around the factory and the Elgin heavy industries, was also the product of spatial-social environmental discrimination, but on the basis of class (or socio-economic status) rather than race. Indeed, one could argue more generally that the most marginalized and vulnerable people in society are more likely to live in areas next to environmental pollution, either through intentional discriminatory policies or through indirect means, such as housing affordability.

The former resident of Glenview-Silvertown, steelworker, and community activist Jim led a campaign to Environment Canada in 1999 to highlight the health impacts of the Cyanamid chemical factory on the residents of Glenview-Silvertown (as discussed in chapter 2). His claims were supported by the stories of cancer, blood diseases, respiratory illnesses, and other health problems from many of his former Glenview-Silvertown neighbours. However, Environment Canada undertook a study into the area and concluded that the health problems of Glenview-Silvertown residents were not related to its physical location, but rather to other factors such as low socio-economic status, higher levels of smoking and alcoholism, or to other historical industries in the area (Pelligrini 2005). One of the key questions in this area is whether the legacies of toxic pollution – of the Cyanamid dust – still linger within the physical environment; whether they exist only in embodied form, in the health problems of residents who lived near the factory during its years of operation; or whether, as sceptics in Environment Canada would argue, they are non-existent in both forms. My research suggests that legacies do linger in environmental and embodied forms, although these are notoriously difficult to measure. On another level, I would argue that the struggle in this socially excluded community is over the issue of corporate accountability and recognition of past and present injuries, a theme which will be explored further in chapter 7.

Walker, Newcastle upon Tyne

Within the city of Newcastle upon Tyne there are wide socio-economic gaps between places of regeneration and redevelopment, which fall primarily in the city centre, and places of social exclusion and stigma, which fall outside of the city centre, such as Walker. In fact, one of the first people who told me about Walker during an early field trip in June 2005 to search for ruins was a friendly middle-aged "Geordie" (Newcastle local) man whom I met in a lively pub in the city centre. He suggested that Walker would be a good place to look for industrial ruins, and gave a colourful description of his own negative impressions of Walker as a blue-collar worker who had grown up in Newcastle. Later on in my research, the Anglican Minister in the East End also highlighted some of these contrasts. He brought me to a church at the top of a hill in Byker (a ward that neighbours Walker and has similar levels of socio-economic deprivation), from which we could see the regeneration along the Newcastle-Gateshead quayside, including the Baltic Contemporary Art Gallery, the Sage Music Centre, and the executive flats. The minister described the contrast between the history of regeneration over the past twenty to thirty years and the situation in deprived communities: "You have this juxtaposition between this particular view of regeneration and culture and communities like Walker and Byker, which increasingly become more and more cut off from the centre of the city" (interview, 10 November 2005). However, as the economic development officer for Newcastle City Council pointed out, the quayside regeneration is a relatively recent development: "You know what it looks like, but you probably don't know what it used to look like. Twenty years ago you would have never gone near the area because it looked like an open sewer, but now you can approach it quite safely" (interview, 30 August 2005). This reflection highlights the idea that "successful" post-industrial transformations of derelict places are possible. However, the wider perspective of post-industrial change across the city also shows the uneven nature of regeneration, where some places are redeveloped (such as city centres and waterfront properties) while others are neglected (such as peripheral old industrial areas).

Despite the wide socio-economic gaps between areas of regeneration and areas of ruination in Newcastle, there are more socio-economic continuities than discontinuities within the context of Walker itself. The social and economic problems that have accompanied the decline in the physical infrastructure of Walker include high unemployment, considerable population loss, high levels of crime, low educational attainment, high teenage pregnancy rates, low housing prices, and health problems (MacDonald 2005). Local residents also flagged other problems such as high drug and alcohol use, incidents of suicide and drug overdose, growing racism with the recent influx of asylum seekers, and incidents of domestic violence against women passed on in cycles through the generations. Some houses in Walker have been abandoned and some have been demolished, with patches of vacant land and housing. On a driving tour of housing developments in Walker with two of my informants, Sheryl and her uncle, in a family-run taxicab, Sheryl pointed out tracts of vacant land where houses had been demolished and buildings that were set to be demolished (figure 5.2). On a walking and driving tour of churches in deprived parts of the Walker community with the Anglican Minister for the East End of Newcastle, we observed that the major shopping centres in Walker, such as Church Walk, had become a fraction of their former size, with few shops and services and meagre healthy and affordable food choices (10 November 2005). The leisure services available at the relatively modern community centre located in the "community focus" area just north of Pottery Bank, were considered good in quality but deemed unaffordable by most local residents (various interviews, 2005–6). Transport links were poor, with the nearest metro station 1.5 kilometres north of Walker, and infrequent bus services. The residential area of Hexham Avenue, which was once a highly desirable area within Walker and housed shipyard workers, was full of run-down and vacant buildings, many of which were slated for demolition under the proposed regeneration. These elements of social and economic deprivation in Walker can also be read as consistent with the industrial past: the social, economic, and cultural legacies of industrial ruination.

I first experienced the landscape of industrial ruination in Walker when I rode the metro one morning in June 2005 to Wallsend (on the northern edge of the Walker industrial riverside, just over the ward boundary) and got off almost on the doorstep of Swan Hunter shipyard. At that time, Swan Hunter was still in operation, although it faced imminent closure and arguably was in a process of ruination. The shipyard building itself was square and grey, with a 1960s feel, and did not seem iconic on first glance. I then walked west towards the city along Hadrian's Way, a pathway that runs next to the industrial riverside, more or less parallel to the main road. Hadrian's Wall became a World Heritage site in 1987, and the local portion of Hadrian's Way was developed as a cycleway and tourist route as part of Walker Riverside regeneration and reclamation schemes in the 1990s, and as the endpoint of the 135 kilometre pathway that follows Hadrian's Wall through the north of England. Along the pathway, I passed numerous warehouses, junk yards, shipyards in various states of abandonment, and very few people, apart from several weary-looking tourists with backpacks heading towards the endpoint of the path in Wallsend. They looked slightly bemused to end their magnificent Roman archaeological journey amidst weeds and industrial abandonment. I had to peer through overgrown trees and bushes that lined the cycle path, and I felt isolated as I walked. There was little traffic on the main roads and no evidence of obvious escapes into consumer society (e.g., cafes, washrooms, shops, services). What struck me most was the enclosed space: the buildings, warehouses, and stripped shipyards were fenced off from public access. As I moved further west, beyond crumbling brick warehouses with broken windows and piles of metal, I came to newer constructions: large vinyl-sided block-shaped buildings which housed offshore companies. Later in my research, I returned to these sites of industrial ruination many times, but it was on this first walking tour that I gained the most enduring

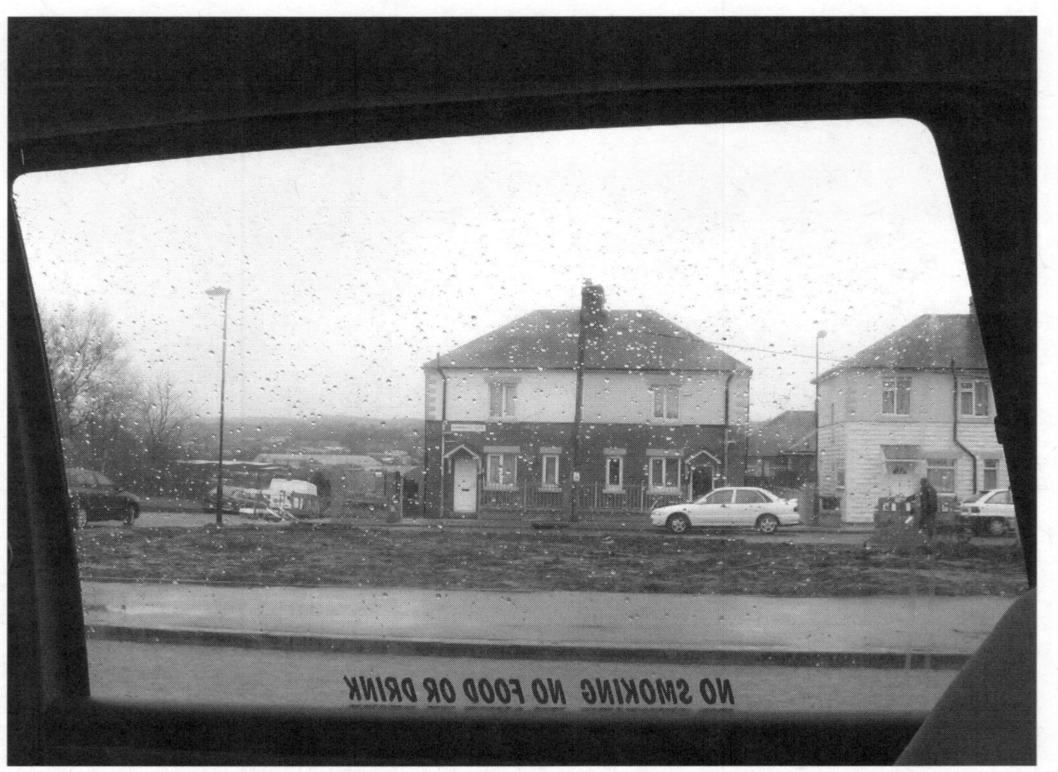

Figure 5.2 View from taxicab of vacant land after demolitions, Walker, March 2006

impressions of the space as separated, contained, regulated, and in various stages of industrial use and disuse.

The social life of the adjacent residential community of Walker has little direct interface with the sites of industrial ruination. Most of the industrial riverside area is blocked off from public access for reasons of environmental safety and security, as I discovered during my numerous attempts to investigate the sites of ruination. The community and industrial areas are divided by a main road, and local residents seldom, if ever, interact with the old industrial sites. The only exceptions are the limited employment in the new or remaining industries, informal (and often illegal) activities, or through walking along Hadrian's Way, as I had done. Although some residents noted that the pathway could make for a pleasant walk because of its view of the river, they expressed concern over its safety and derelict state. Echoing my own observations, the residents said that the area is overgrown with bushes and poorly kept, with limited access to roads and services. Teenagers often drink there at night, and hypodermic needles have been found along the path. One resident said: "I used to take my dog along there every night to meet my partner, because he cycles from work and he used to come along the bottom road, but it's terrifying. I wouldn't go down there at night ... The bushes are really overgrown ... 'Cause it is a nice walk if it was all open" (interview, 21 March 2006). Another resident added: "You get teenagers starting to hang around down there 'cause there's nothing for teenagers to do. They're hanging around down the banks drinking and what have you. It's a bit intimidating" (interview, 21 March 2006).

The spatial divide between Walker's riverside industries and community is reflected in a policy gap between housing-led regeneration plans for the community and an ambiguous plan to attract offshore industries and other economic development along the riverside. The manager for Newcastle City Council's regeneration development partner noted this gap (interview, 2 December 2005). The manager also admitted that the problem of creating employment in the area was a weak point in regeneration schemes, and that the skills requirement is too high for the Walker community to benefit from the few jobs available on the riverside.

Many other interviewees also expressed the view that the limited employment opportunities on the riverside do not match the skills of most people living in the Walker community. Furthermore, most of those workers who have the training necessary for the work available on the riverside are elderly. Skills training for shipbuilding on the River Tyne stopped altogether in the 1990s. During its last years of operation, Swan Hunter took on apprentices, although the trained workers subsequently had to move elsewhere to use their skills. According to some interviewees, there is some translation between skills in shipbuilding and the skills required to work in offshore oil and gas industries, but not usually in the higher-paid jobs. Moreover, the types of industries that remained on the industrial riverside during my research did not have similar associations with pride: in the years leading up to the end of shipbuilding on the Tyne, shipyard workers spent their hours on decommissioning rather than building. However, since 2009, Shepherd Offshore, one of the main offshore companies located on Walker Riverside, has invested heavily in offshore wind turbines. This development has the potential to bring a new source of pride for work on the Tyne, although these jobs would require very high skill levels, which do not match the local skills in the area (several follow-up interviews, July 2009).

Since the time of my fieldwork, there has been one notable attempt to address the local skills gap. In 2007, Building Futures East, a voluntary sector organization, was established in Walker's former shipbuilding area with the aim of regenerating deprived North East communities, particularly Walker, through providing free training, education, and development opportunities to vulnerable individuals. The vision of Building Futures East contrasts with past practices, which have tended to stigmatize and punish unemployed people and people in receipt of benefit, particularly in areas with high levels of social and economic deprivation. However, while this initiative is a positive step towards helping the most in-need young people in the area, it is not comprehensive. There are still very few local jobs, high levels of socio-economic deprivation, and mounting housing problems related to the failures of regeneration, as will be discussed in the next chapter. There are strong policy and skills gaps in Walker, which echo the spatial gaps between the community and the riverside. This resembles the picture of the hourglass economy of income distribution (Bluestone and Harrison 1982), with a large gap in jobs that produce middle incomes. Although these gaps are local, in that they are contained within the geography of Walker rather than set in relation to the rest of Newcastle, they emphasize the contrast between insiders and outsiders. Much of the existing local employment does not match the local skill set, so the jobs go to non-locals. The spatial division between industry and community reflects a shift between an era of growth, production, and physical flow to one of unemployment, stagnation, and disconnection.

The landscape of industrial ruination in Walker Riverside is separated from the adjacent community through physical barriers, the decline of viable employment, and the gaps in policy focus and skills between the riverside and the community. Nonetheless, and perhaps at a more fundamental level, the landscapes are deeply interconnected. The socio-economic and physical deterioration of the Walker community has mirrored industrial decline. The interconnections between landscapes of industrial ruination and landscapes of community suggest an uneven geography of capitalist development. Walker has remained isolated from the rest of the city, stigmatized as a place of socio-economic deprivation, and resistant to top-down forms of planned regeneration that seek to transform it into a middle-class commuter riverside village. Regeneration plans have been primarily property-led, based on the demolition of existing houses and the development of higher value riverside properties. But they have not addressed the lack of jobs or the loss of skills in the area, and it is difficult to see how Walker will overcome its spatialized socio-economic deprivation and exclusion.

Ivanovo

The city of Ivanovo contrasts with the cities of Niagara Falls and Newcastle upon Tyne, for industrial ruination in Ivanovo is embedded within the cityscape, rather than concentrated in particular areas of the city, and most of it is only partial. When I first arrived in Ivanovo, I was struck by the implications of living within a city full of ruins, which was when I first thought critically about insider and outsider perspectives. In my field notes, I wrote:

> Reflections on distance and proximity: Industrial ruins are only "spectacular" and "sublime" from a distance. For example, as the "Way to Russia" guide suggests, it's best to view the Soviet ruins as one passes through, rather than stopping too long. I felt almost as soon as I arrived the social reality (stark, depressing, mundane) of living amongst such ruins. To call the ruins aesthetically beautiful is already to put oneself at a distance. It is a privileged position. (7 September 2006)

The landscape of partial industrial ruination blends into the cityscape of Ivanovo, with both spatial and socio-economic manifestations. This is partly related to the historical blending of industrial and Soviet activities and identities within the city. There are in fact two types of ruins visible in Ivanovo: industrial ruins, in the form of abandoned and semi-abandoned factories, and Soviet ruins, in the form of dilapidated Soviet murals, buildings, and monuments.

The historical textile and Soviet identities of Ivanovo are present within the cityscape. The industrial city of Ivanovo was created during the Soviet period, which had interesting implications for its design and collective identity. The weaving villages of Ivanovo and Voznesensk merged in 1871, but the rapidly expanding textile conurbation only achieved city status in 1918 and was renamed Ivanovo in 1932. In 1905, Russia's first Soviet of Workers' Deputies was created in Ivanovo, which remains a point of pride for the city today. The Park of 1905 is situated near the Krasnaya Talka factory, along the River Talka, where the first Soviet of Workers' Deputies was created. This park contains a large monument to the revolutionaries of 1905 and statues of local communist heroes. In Moscow and St Petersburg, many Soviet street names have been changed, and selected monuments have been destroyed. However, there has not been a similar move away from the past in Ivanovo, where Soviet references are embedded within the map of the city. The largest streets are named after Marx, Engels, and Lenin, and smaller streets are named after local communists. Revolution Square, located in the centre of the city, marks Ivanovo's historically significant place as "the third Russian proletarian capital after Leningrad and Moscow." There are Soviet murals on numerous buildings, including the central train and bus stations, apartment blocks, and factory walls. Several prominent statues of Lenin remain in the city centre.

I asked Ivanovo residents whether there had been any political move in recent years towards renaming streets or tearing down monuments. None of my interviewees recalled any such discussions, although several remarked that there had been some discussion of changing the name of the city back to Ivanovo-Voznesensk.[3] Perhaps the total collapse of industry in the mid-1990s and the accompanying food and housing shortages seemed like more pressing issues during those years of sudden change – and perhaps this attitude, of working around and living with what was there, was also a reflection of a utilitarian approach in relating to space.

In addition to industrial and Soviet ruins, Ivanovo contains a number of abandoned or derelict residential and commercial buildings. As I observed on one of my strolls through the city, "Friedrich Engels Street, a wide and busy street which runs parallel to the Prospect Lenina, is rather

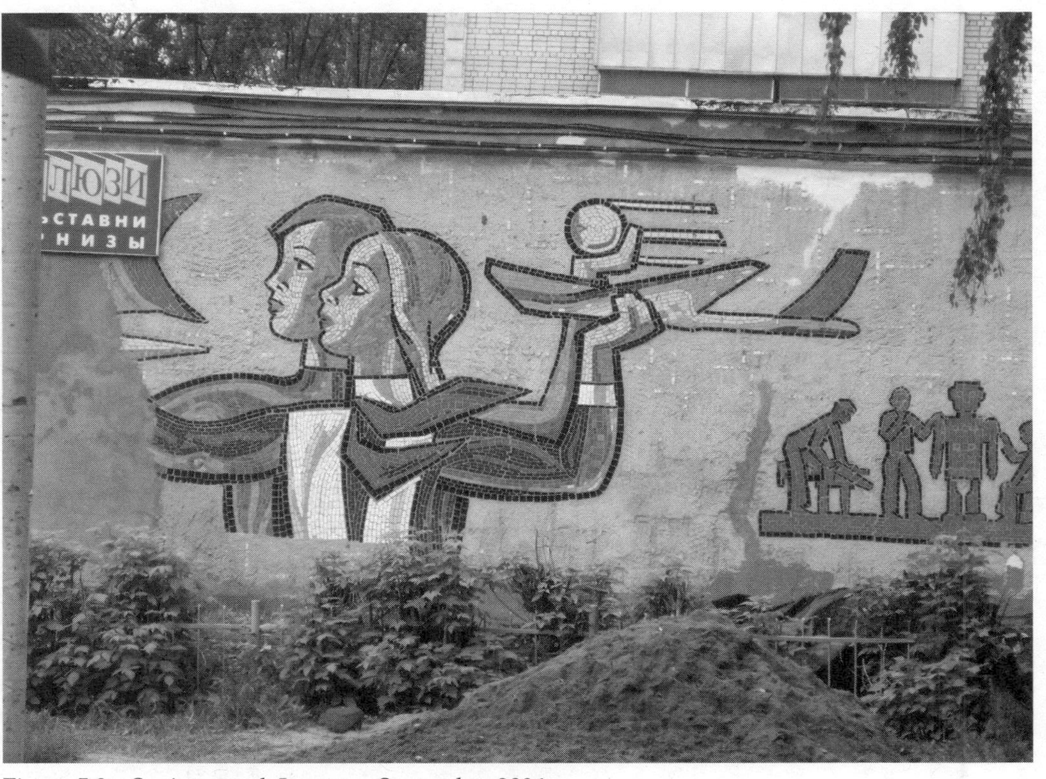

Figure 5.3 Soviet mural, Ivanovo, September 2006

grim – some dilapidated concrete buildings, residential blocks, and some random shops selling light fixtures, furniture, and produce" (field notes, 6 September 2006). I also noticed that city roads were poorly paved, and that many roads, including the one in front of my accommodation, were made of dirt. Yet city traffic in Ivanovo was heavy, even along unpaved roads, and old cars and *marshrutkas* crowded and polluted the streets. During my field research, particularly on rainy days, the streetscape of Ivanovo seemed to be filled with a gritty atmosphere of industrial decline which was reminiscent of descriptions of Manchester or Newcastle in the 1970s or 1980s. I observed many people drinking alcohol openly in the streets, in shops, in parks, and along the river at all times of day. Although this is customary in Russia, it seemed more pronounced than in St Petersburg, where I had spent a month prior to my arrival in Ivanovo.

The cityscape of Ivanovo has undergone a great deal of change since the end of the Soviet Union, but not (at least during the time of my field research) in the direction of the post-industrial or the post-Soviet. Rather, many of the changes are marked by the overwhelming presence of ruination: abandoned factories and dilapidated residential and commercial buildings. Sitar and Sverdlov (2004) suggest that the urban picture in Ivanovo reflects overlapping political eras in urban planning, as various styles of Soviet construction – *stalinskie doma* (Stalin houses), *krushchevki*, and *brezhnevki* – have each left their imprints on the city. However, none of these styles was fully realized, leaving the city with an inconsistent architectural character (Sitar and Sverdlov 2004, 9–10). Ivanovo has engaged in piecemeal, market-oriented renovations of derelict properties, and some Western-style shops that offer a range of products, such as mobile phones, have opened in the city centre. However, at the time of my research, many of these shops had already been abandoned due to economic difficulties. For example, the Plaza, a former theatre which was abandoned for over thirty years, was converted into a shopping centre in the 1990s and closed down in August 2006, reportedly because the building did not meet health and safety regulations.

The landscape of ruination within the cityscape of Ivanovo reflects relatively recent and widespread industrial decline. The abandoned, mostly abandoned, and semi-abandoned textile factories in the city centre are a spatial and scalar testament to the importance of the local textile industry. The Soviet street names, monuments, statues, and crumbling mosaics, and the dilapidated residential and old industrial buildings, suggest a city which is living and functioning among its own ruins. Many interviewees suggested that the city is gradually recuperating, with more green spaces, less rubbish in the streets, and economic improvements. Textile factories that seem abandoned continue to operate through political barter exchanges and the tenacity of workers and employers in maintaining the ideal of the "Russian Manchester." Accounts of socio-economic deprivation in Ivanovo suggest that there are strong continuities between the landscape of ruination and the socio-economic landscape, and these continuities represent industrial legacies of decline and hardship. Yet the prevailing legacy of industrial ruination is perhaps less centrally about socio-economic deprivation and despair, although that is certainly part of the story. Rather, the industrial legacies in Ivanovo are evident in its enduring textile identity as the "Russian Manchester," and in the practical ways that people live their everyday lives in and among the ruins.

If the landscape lacks juxtapositions and discontinuities such as those found in Walker, does Ivanovo fit into an uneven geography of capital development, or is such a concept irrelevant? Ivanovo relates to an uneven geography of capital redevelopment precisely through its exclusion from the capital redevelopment that has occurred in more affluent places throughout Russia and the rest of the world. In fact, this is not dissimilar to the discussion of Newcastle in relation to the rest of England, and of Walker in relation to Newcastle. In Walker, juxtapositions within the community itself are related mainly to a division between insiders and outsiders within the community. In Ivanovo, outsiders are relatively rare, but the discrepancies between insider and outsider perspectives are significant, and will be addressed in the next chapter. The main theme that can be drawn in this discussion of the relationship between the landscapes of industrial ruination, of Soviet ruination, and of the cityscape as a whole, is that these landscapes are closely intertwined, and there are more continuities than discontinuities with the industrial and Soviet past. It is difficult to separate Soviet history from industrial history in Ivanovo, where it is even difficult to separate the past from the present.

Conclusion

This chapter has combined a range of visual, spatial, mobile, and ethnographic methods to read social and spatial landscapes of industrial ruination as a lived process that is linked to wider socio-economic processes of deindustrialization, an uneven geography of capitalist development, and heavy socio-economic burdens for adjacent urban and community spaces. This analytical reading is linked to approaches within urban sociology, human geography, and contemporary archaeology which interpret places as palimpsests, concerned with unpacking their multiple layers and meanings. First, I analysed the landscapes of industrial ruination, reading and interpreting clues about the scale, scope, nature, geography, and process of deindustrialization. Second, I analysed how the abandoned factories and shipyards related to the wider landscapes of adjacent residential and urban environments. The process of reading landscapes reveals the deep interconnections between spatial, socio-economic, and temporal layers. These entanglements are important for understanding the wider processes and complexities of industrial ruination.

In each case, there were socio-economic continuities between industrial and urban decline, with clustering and juxtapositions evident in the uneven geographies of capital abandonment. In Niagara Falls and Newcastle upon Tyne, there were sharp contrasts between different areas of the cities. In Ivanovo, the landscape of (partial) ruination is interwoven with the cityscape rather than clustered in a particular neighbourhood, with a unique fusion of Soviet and industrial ruins, but the city-region as a whole is marginalized and stigmatized in relation to the rest of European Russia. There were also distinctive themes in each case. In the Highland and Glenview-Silvertown communities of Niagara Falls, the theme of discriminatory spatial segregation or exclusion of people is linked to class and race. In Walker, despite socio-economic continuities between industrial and urban decline at the local level, there are considerable gaps between the industrial riverside and the neighbouring community, both physically, through regulation, and socio-economically, with few employment opportunities and local skills that do not match the existing highly skilled jobs in offshore technologies.

The spatiality of socio-economic deprivation and exclusion was a common theme across these cases, despite local differences and patterns. This relates to Harvey's (1999) notion of an uneven geography of capitalist development, and to Madanipour's (1998) notion of the spatiality of social exclusion. It also highlights the fact that landscapes of industrial ruination are rarely truly abandoned. Photographic representations of industrial ruins tend to focus only on the physical landscapes of abandonment, rather than investigating the social life that surrounds them. People are reluctant to leave communities steeped with memories and relationships, even when jobs have disappeared, infrastructure is crumbling, and the land is full of toxic contamination. I will explore this theme of place attachment in areas of industrial ruination in the next chapter.

Chapter Six

Devastation, but also Home

Despite socio-economic deprivation and material devastation in areas of industrial decline, houses, neighbourhoods, and cities can become invested with notions of family and community unity, nostalgia for a shared industrial past, and stability amidst socio-economic change. This chapter examines place attachment to "home" in the context of disruptive post-industrial change, where homes and communities have come under threat from demolition, stigmatization, or toxic contamination. The theme of place attachment emerged in three case studies of the landscapes of industrial ruination: Highland, Niagara Falls, New York; Walker, Newcastle upon Tyne; and Ivanovo, Russia. I argue that narratives of conflicted place attachment – of devastation, but also home – reveal some of the contradictions and uncertainties of living through difficult processes of social and economic change.

The concept of place attachment in areas of industrial decline raises some key questions. In the economic context of low employment opportunities, physical dereliction, and trends of depopulation, why do people still attach meaning to their homes and communities? What is the value of place attachment to individuals, families, and communities? What is lost when people have to be mobile? This research suggests that neither mobility nor fixity creates a sense of loss, but that limited choice – based both on economic structures and on conflicted feelings of place attachment and despair over economic realities of industrial decline – does. In the cases of Walker and Ivanovo, the primary barrier to redevelopment was people's attachment to their homes, while in the case of Highland, the primary barriers were existing negative environmental and economic factors in the area (close proximity to heavily contaminated abandoned industrial sites), regardless of people's local attachments. An analysis of place attachment in areas of industrial decline draws attention to the social and psychological impacts of uncertainty, disruption, and stress of lived experience through difficult economic transition.

Place Attachment, Community, and Home

There are a number of different definitions of place attachment within the literature, although most definitions emphasize some form of an affective bond between people and landscape. Low (1992, 165) provides a useful distinction between cultural and psychological definitions of place attachment: "Place attachment is the symbolic relationship formed by people giving cultural shared emotional/affective meanings to a particular space or piece of land that provides the basis for the individual's and group's understanding of and relation to the environment." The concept of place attachment used in this chapter follows this cultural definition but goes further: place attachment is related to social and economic processes, and place itself is not simply a "particular space or piece of land" but rather, it is "inhabited," similarly to the notion of landscape as "dwelling" (Ingold 2000).

The phenomenon of place attachment has been addressed mainly within two contexts, most notably within the environmental psychology literature. The first context relates to the psychological effects of residential mobility, particularly in cases of forced relocation. In a classic study of a working-class area in the West End of Boston, Fried (1963) argues that residents had strong place attachment to their community despite the poor quality of residential housing. He shows that there is not necessarily a decrease in place attachment in areas of urban decline. Moreover, he argues that the psychological effects of relocation were disruptive for the West End residents' sense of continuity in life, and that residents continued to feel place attachment even after relocation. Other scholars also argue that there are negative psychological effects associated with residential mobility, including a sense of being rootless and placeless, and a lack of continuity and social cohesion (Altman and Low 1992; Relph 1976; Tuan 1977). However, recent scholars contest the negative psychological effects of mobility. They argue that there can be positive effects of both mobility (routes) and place attachment (roots) which are not necessarily opposed but intertwined (Clifford 1997; Gilroy 1993; Gustafson 2001).

The second context in which place attachment has been studied is related more directly to community studies and community development in areas of urban decline. Fried (2000, 193) has extended his analysis of place attachment in poor communities to make a more general claim: "Attachment to place is a characteristic feature of life in many poor, ethnic, immigrant communities. The development of a sense of spatial identity is a critical component of attachment experiences in such local areas." Some scholars argue that place attachment in depressed neighbourhoods might function as a form of social and community cohesion that could be used as a policy tool in neighbourhood revitalization (Brown et al. 2003). As Fried suggests, the phenomenon of place attachment in areas of industrial decline is neither surprising nor novel in itself. It can also be explained by the psychological phenomenon of "adaptive preferences," whereby people adapt to difficult living circumstances as a strategy for coping and survival. It is worth paying attention to Sen's (2009, 283–4) critique of arguments that justify continued deprivation and poverty around the world because of people's adaptive preferences in coping with difficult life circumstances:

> The adaptive phenomenon particularly affects the reliability of interpersonal comparisons of utilities, by tending to downplay the assessment of the hardship of the chronically deprived, because the small breaks in which they try to take pleasure tend to reduce their mental distress without removing – or even substantially reducing – the actual deprivations that characterize their impoverished lives. To overlook the intensity of their disadvantage merely because of their ability to build a little joy in their lives is hardly a good way of achieving an adequate understanding of the demands of social justice.

Adaptive preferences partially explain the apparent contradiction of place attachment in areas of industrial decline. However, the context of post-industrial transformation in these cases introduces an additional dimension of uncertainty over the threat of change. This research focuses on place attachment as a complex legacy of industrial ruination that is reflective of lived experiences.

Both Walker and Highland represent specific communities or residential neighbourhoods adjacent to sites of industrial ruination, whereas in Ivanovo, industrial ruination is spread throughout the city. Studies of particular communities and neighbourhoods have a long and varied history of scholarship, and there have been a number of classic studies of disadvantaged (working class; ethnically segregated) communities and neighbourhoods in particular (cf. Bell and Newby 1971; Winson and Leach

2002; Young and Willmott 1957). The concept of community has been criticized because of its relationship to romanticized and nostalgic notions of social cohesion and place identity, and its tendency to represent "neighbourhoods as relatively class-homogenous, small-scale, easily delineated areas with clear borders, hosting relatively cohesive communities" (Blokland 2001, 268). While I use the concept of community within my research, I recognize that it is a contested term with political connotations.

The concept of "home" is closely wrapped up with that of community, and also carries contested and contradictory meanings. Mallett (2004, 65) argues that home is a multi-dimensional concept, asking: "Is home (a) place(s), (a) space(s), feeling(s), practices, and/or an active state of being in the world? Home is variously described as conflated with or related to house, family, haven, self, gender, and journeying." Mallett concludes that the experience and study of home relates to all of these themes, and that research on it can be value-laden depending on different contexts and motivations. For example, the home is a political terrain of struggle over gender roles and domestic expectations; the conflation of house and home encourages a capitalist culture premised on home ownership; and the association between family and home relates to conservative ideologies about preserving traditional nuclear family values (Mallett 2004). Recent literature on the material culture of the home has focused on interior design, home possessions, mantelpieces, photographs, radios, and other contents of the home (Hurdley 2006; Januarius 2009; Miller 2006). Much of this literature focuses on the construction of home and personal identity through consumption. By contrast, this research examines homes in a different context: homes under threat of demolition, deterioration, dislocation, and/or contamination. These are not new homes, and their role in the construction of personal identities is largely in the past.

Place identity, as distinct from individual identity, is an important concept in considering the notion of home within the context of this research. The concept of place identity has been explored within the field of human geography to discuss a range of issues around people's attachments, memories, and identities surrounding place (Cresswell 1996; Gustafson 2001; Massey 1994; McDowell 1999; Relph 1976; Tuan 1977). Tuan was among the first theorists to discuss the affective bond between people and place; he termed this "topophilia." Cresswell (2004) argues that for many, the idea of home is the most familiar example of place and its significance to people, and he connects the centrality of home in humanistic and phenomenological approaches to place with Heidegger's focus on "dwelling" as the ideal kind of "authentic existence." Feminist geographers such as Gillian Rose (1993) criticize romantic notions of home as an ideal place: "[Tuan's] enthusiasm for home and for what is associated with the domestic, in the context of the erasure of women from humanistic studies, suggests to me that humanistic geographers are working with a masculinist notion of home/place" (Rose 1993, 53). Massey (1994, 119) argues that there is "no single simple 'authenticity' – a unique eternal truth of an (actual or imagined/remembered) place or home – to be used as a reference either now or in the past." In other words, place and home identities are never stable, fixed, or bounded, but they are always constructed.

In each of the cases, home represented different notions which were entangled with living memories of industrial ruination. In Highland, homes represented oases of achievement and family endurance amidst hazardous contamination and severe poverty. In Walker, the importance of homes as sites of family and community stability was revealed through the threat of demolition. In Ivanovo, some residents in dilapidated traditional Russian houses also faced that threat, but in general they were attached to their city as a whole rather than to particular communities within it.

Highland, Niagara Falls, New York: Family, Church, and Community

Place attachment to home emerged in the Highland community as a symbol of family, church, and community within an intensely socially and racially excluded community located in close proximity to contaminated industrial sites. According to various Highland residents, community solidarity was very strong, and they perceived their community as one big family. That solidarity came from strong cultural ties from a shared heritage from the South; the shared experience of living with spatial, socio-economic, and racial segregation from the rest of the city; and from extreme poverty, joblessness, health problems, and contaminated land.

According to Carl, an African American former resident (see also chapters 2 and 5), there is a discrepancy between the way that neighbourhoods are perceived, the way that they are experienced, and the social and economic realities of life within them (interview, 23 March 2007). He said that the rest of Niagara Falls perceives Highland only as an area with the largest concentration of low income families and with high rates of crime. He also argued that this negative image came from the media portrayal of the community, which often connected stories of crime in other areas of the city back to Highland. His own view of the area contrasted with this stigmatized perception:

> For the families that live here, it's a very different sense of what the community is all about. It's a stronger sense of home ... I think that the people that are here have a very strong sense of community among themselves, and I think that almost the poorer they are, there may be a stronger sense of community because they kind of watch out for each other. People will get sick and neighbours will bring them food and make sure that they have something to eat. And then maybe that sense that I have to help this person, because if something happens to me and my family's now all gone, who's going to be there? You know, it's just going to be the neighbours that are left. (interview, 23 March 2007)

During our driving tour of the Highland area, Carl described the history of different homes: who lived in each house, whether they had undertaken home improvements, and when people had died or moved. He also pointed out how close the chemical factories were to people's backyards. His knowledge of the community and the streets was intimate and reflected a rich local knowledge, but he also situated that knowledge in relationship to the city as a whole.

One of the clearest examples of place attachment to home in Highland is evident in the narrative of Mary, the elderly African American resident and Baptist minister discussed in chapter 2. Regardless of the social and economic problems of crime, drugs, poverty, and poor health in the area, she was very positive about her experience of living in a community with close friends and family. She said that she has had a good life in Niagara Falls despite the economic hardships because of family, church, and community. She spoke of the individual homes and families on each street with knowledge of the details of each home and family, and a sense of each street as a unique space. I was struck by the sense of place that Mary evoked in her descriptions of Virginia Avenue:

> But as far as Virginia Avenue runs, there wasn't [sic] when I came here, there was just one, two, three, four, five, maybe six houses here. And that's all. The tall house when you come up around the curve around there, that house was there, and the [family's] house was there, and they had this other house, that light green and dark green house, and then the house next door to me, and the [family's] house, and there was another house up the street there.

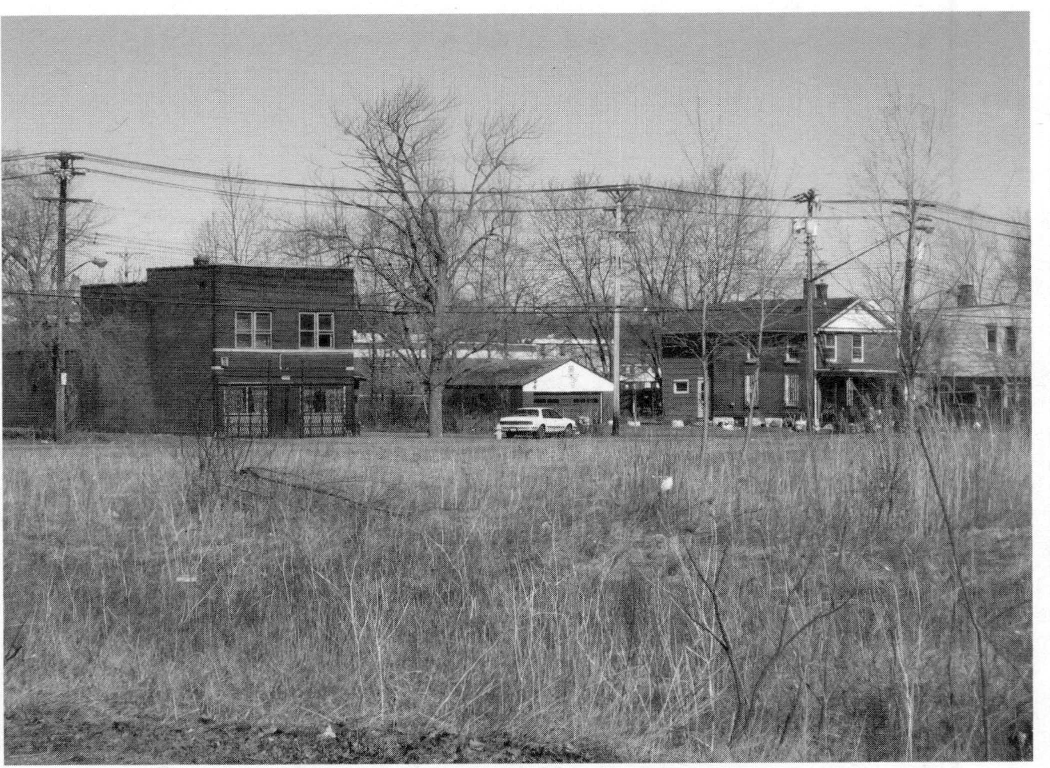

Figure 6.1 Highland Avenue residences and overgrown brownfield, April 2007

She also described the social world as one of a big extended family:

> But as far as Virginia Avenue, we've had a neighbourhood, one thing, a very good friendly neighbourhood. We had on Virginia Avenue a black club that we was very close with one another. We visited with one another, you know, and had good times in each other's homes. We would have the meetings, and it was like a social club 'cause each family would prepare a nice big meal and we would have our meetings and then we would eat and unite and we was served, and we was like a family here. We supported each other, the Virginia family; we came together and we supported each other. And it's been good living here for me. And my health isn't that good but I'm doing okay.

Mary did not express much interest in whether the former chemical factories in Highland were contaminated or not, or in whether they related to the local health problems. She seemed reluctant to equate any negative experiences with the Highland community itself and instead blamed the closure of the factories for decline and deprivation in the community. She emphasized the strong sense of family and community in Highland and the importance of the churches, which she said work together and are like families too.

Many people in Highland linked the importance of home to the concept of family. Several interviewees described the difficulties that families endured in order to afford to build homes and support themselves. One interviewee explained this attachment as follows: "One of the things that I think is sort of difficult about this city is that the families are families that their homes is sort of all that they have that's left, and no matter what, they don't want to leave their home, they don't want to sell their home" (interview, 23 March 2007).

People referred to the structure of families in the past, in which a man worked in the factory and supported a wife and children, with a sense of nostalgia and loss. They lamented the demise of this structure, as it was no longer tenable without the possibility of men gaining full-time, high-wage employment. Alan, another African American Highland resident, commented on the change in the community and the erosion of the traditional family home:

> We have about a third of the homes [in the Highland community] that we had when the city was booming. Most of those homes have been torn down burned out and torn down. Of the remaining homes, you only have a third

> of the disposable income that you had. The house that I live in, and I live there alone, used to be a two family home. There were nine people living in that house at one time, now there's one. But you look down the neighbourhood, you still see, there's a house there. You can no longer make the assumption that there's a family there. (interview, 21 March 2007)

Carl echoed Alan's thoughts about the impact of the loss of manufacturing jobs on families:

> You can find a job that's part-time, that's paying just enough that you can survive, but not really enough to do anything, not enough to really save. Not enough to really feel secure financially and make plans, you know. You hear people say, well, you know, "I can't afford to get married; I can't afford to do it. Because I'm barely able to take care of myself and I can't take on a wife and children. I can't buy a house." (interview, 23 March 2007)

Many local people seemed to wish for a return to the traditional Fordist structure associated with the post-war golden era of high-wage production-line manufacturing and nuclear families. However, they were not disdainful of elements of what has been theorized as the post-Fordist social and economic structure (cf. Amin 1994; Jessop 1991), which has different gender roles, forms of employment, and consumption patterns. In fact, the Fordist structure was the only one they had experience with, since nothing had replaced the logic and socio-economic reality of the heavy industrial era, as the knowledge and service economy has in other cities around the westernized world.

These reflections on the decline of families demonstrate the salience of memories and perspectives which are related not only to stable, high-wage manufacturing employment, but also to the social world views associated with the Fordist era more generally. They suggest that place memory of former industrial communities is not only related to class, but it is also related to some extent to gender and gender roles. The social and economic aspects of the post-industrial paradigm – the growth of the service sector, the promises of knowledge and clean technology, and the feminization of labour – all bypassed Niagara Falls during its years of deindustrialization.

Despite community revitalization plans in Highland, the family home, the church, and the community continue to exist as havens of solidarity amidst decline, sickness, poverty, and contamination. In this case, place attachment is not sufficient as a policy tool for community renewal, as

Brown, Perkins et al. (2003) have suggested, because the costs of clean-up, let alone economic revitalization, are too high. The lack of economic resources as an obstacle to community redevelopment is particularly evident when framed in the wider context of industrial decline, depopulation, and contamination in Niagara Falls, New York. Carl remarked that people in Highland would build solid and attractive homes despite the fact that they could only resell them, if they could resell them at all, at a fraction of what they cost to build (interview, 23 March 2007) – a phenomenon that is a result of residents' long-standing attachment to their community. People's homes, and their networks of community, were not under threat of demolition, but their livelihoods, their business district, their health, and everything else had gone.

Walker, Newcastle upon Tyne: Home and Community under Threat

Walker is a traditional working-class community that formed around the shipyards and industries of Walker Riverside in the 1930s, and almost no new building occurred after the area went into decline in the 1970s. In 2006, 70 per cent of the total housing stock consisted of local authority (social) housing. The city council's regeneration plans for the area were based on large-scale demolitions of these types of houses, which the city thought symbolized deprivation and decline, even though many of them were solidly built. The primary goal of regeneration was to attract middle-class populations into the area through building attractive riverside housing, which was effectively a form of planned gentrification. The city proposed that displaced residents move into different, temporary housing, with the option to buy or rent the new houses that would be built in place of the old. However, the council was set on a mixture of 80 per cent ownership and 20 per cent rental tenure, and the prices were beyond most residents' means. My research focuses on residents' place attachment to their homes and to their community as a whole during the difficult early phases of regeneration, and particularly looks at the homes under threat of demolition in Pottery Bank, a key area for development because of its riverside location.

In many ways, the structure of community and family networks in Walker reflect patterns similar to Young and Willmott's (1957) study of Bethnal Green, a (then) white working-class neighbourhood in East London. Walker, too, is a traditional white working-class community with roots in an industrial past. Extended families often live within blocks of one another, and family and community are important for most residents. In regards to Bethnal Green, Young and Willmott (1957, 186) conclude that place attachment was related to family and kinship ties: "The view that we have formed and tested more or less daily for three years is that very few people wish to leave the East End. They are attached to Mum and Dad, to the markets, to the pubs and settlements, to Club Row and the London Hospital." One could draw a similar conclusion about the importance of family and kinship to place attachment in Walker. However, in Walker, place attachment can also be framed as a yearning for stability and continuity amidst disruptive social and economic change (as Fried, 1963, argues), and as an idealized vision of community projected onto a turbulent social and economic reality.

During a walk around the area, Sheryl, a local resident, mother of two, and worker in the community and voluntary sector, illustrated the close community and family networks in Walker: she knew the people, places, politics, and local history in intimate detail, and she showed me her house and the houses of her mother, grandmother, and brother, all within a couple of blocks (21 March 2006). Another interviewee, a resident of Pottery Bank whose house was under threat of demolition, said that her father, mother, and children all lived within doors of one another, and thus the impact of demolition and relocation would be "soul-destroying" because it would separate the family (interview, 12 September 2005).

Walker residents expressed particularly strong attachments to their physical homes and to the ideas of "home" and "community." In 2001, the strength of this attachment prevented the original Newcastle City Council regeneration plan, Going for Growth, from going ahead as planned. Going for Growth involved significant housing demolitions within the community, and the residents whose houses faced demolition were strongly opposed to the plan. Although the houses were built as council homes, many of the residents had worked hard to buy their homes and had decorated and refurbished them extensively. It took several years and numerous "community engagements" and "community consultations" before the city council was able to advance with the regeneration plans at the beginning of 2006. A Labour councillor for Walker noted that residents were hostile toward the proposed demolition, particularly in Pottery Bank, which would be a prime location for the development of modern riverside flats (interview, 30 August 2005).[1] The 1930s houses would be demolished and replaced, rather than refurbished, and residents would be relocated within the Walker area.

Newcastle City Council set up the Cambrian show homes (figure 6.2) as examples of new housing for the area in general, and intended to use them to not only to replace demolished homes, but also to attract newcomers to the area. The Cambrian area already had vacant land and hard-to-let properties, so the city council had deemed it an appropriate place to begin regeneration. However, few local residents would be able to afford to move back into the area once it had undergone regeneration and redevelopment. The Labour councillor described the Cambrian show homes as "new eco-friendly cardboard boxes" that were "small, cramped, flimsy" and would likely last only twenty years, and several residents echoed this attitude. One argued:

> We don't want to be taken out of the community, and we certainly do not want the monstrosities they have got on Church Street [the Cambrian homes]. We don't want those buildings, we don't want them shoeboxes, we've got good strong family homes and we want to keep them. We don't want them shoeboxes that you can't do nothing with, you can't swing cats in. (interview, Pottery Bank resident, 12 September 2005)

Since its initial failed attempts at regeneration in 2001, the city council had conducted a series of consultations with the community in Walker in order to obtain the consent of the population. However, the city council delayed a decision about whether the homes would be demolished or not until 2006, when they approved the "third option" in the plan, which called for the highest number of proposed demolitions and significant change for the community. One Pottery Bank resident expressed frustration with the experience of waiting for an answer from the city council about potential demolitions:

> All we want to know is a straightforward answer, yes or no, are they standing up or are they coming down? ... I jokingly said, didn't I, last year that if we haven't got an answer by November [2005], I says I'm going to take part in legal history because I would take Newcastle City Council to court for stress, for compensation, for payment, for stress, for nerves, for sleepless nights. For all the residents: it's mental cruelty; it's psychological cruelty. (interview, Pottery Bank resident, 12 September 2005)

This resident underlined her own perceived lack of voice within the process of regeneration in the community by presenting the idea of taking legal action over the psychological impacts of living with uncertainty as a "joke." However, her point about mental and psychological cruelty was a profound one: several residents reported that the uncertainty of

Figure 6.2 Cambrian show homes, Walker, March 2006

the future of their homes impacted their health and led to stress, anxiety, depression poor mental health, panic attacks, and a feeling of inability to do anything with their homes – sell them, re-paint them, furnish them, or anything at all (interviews, Pottery Bank residents, 12 September 2005). An elderly woman at the community centre in Pottery Bank, who had lived in one of the homes set to be demolished for thirty-one years, expressed a similar view: "People there won't let me know when, how long's it gonna be before they're coming. But they said they're supposed to build houses on Pottery Bank before they pull them down and I'm hoping they're going to keep that promise" (interview, 22 March 2006). Her story is filled with a sense of resignation in the face of change.

The residents were not only concerned with their own homes and stories; they were also concerned about their neighbours. One resident reflected on the importance of the home as an enduring feature of one's life, particularly in the context of old age:

> I know a couple, they got married when they were eighteen, and that was their first house. One of them, the oldest one was eighty-nine about three years ago, and they've lived in that house all their lives, brought their daughter up, and they don't want to go. All their faculties about them, you know, old people, you expect to end your days in that home. It's not just a house, it's not just an estate, it's a home. And it's all their life, past, present, and future. You can't do that to people. (interview, 12 September 2005)

In *The Corrosion of Character: The Personal Consequences of Work in the New Capitalism*, Sennett (1998) highlights the difficult personal experiences of working in the uncertain and risk-laden context of the new flexible knowledge-based economy, arguing that the lack of predictability, loyalty, and long-term relationships between workers and employers makes it difficult to maintain a defining narrative of one's working life. The loss of a home through post-industrial change also reflects a break in the narrative of one's life, a narrative rooted in expectations about family life, community life, and the process of ageing.

Since 2000, the question of housing in Walker has also been significant in debates about the government's "hard-to-let" policy to disperse asylum seekers in deprived areas (discussed in chapter 3). Between 2006 and 2009, particularly after the housing demolitions took place, there were mounting racial tensions over housing resources in the community (five follow-up interviews, 8–9 July 2009). Many of the asylum-seekers had been granted refugee status and were competing with local residents for limited social housing. The shortage led to a crisis of overcrowding in Walker, forcing many people to live with their extended families in two or three-bedroom houses. One resident (interview, 9 July 2009) described the situation as follows:

> I know my friend, she's been looking for a house for twelve months, and she lives in a house with only two bedrooms. She's got three children, her sister lives there and her sister's boyfriend as well, her older sister lives there and the dad, and her and her three kids in a two-bedroom house, and they can't get another one, they can't get re-housed, so she's just stuck really. So they're sleeping on the settees and making do with it because they have to.

These practices within the traditional white community in Walker – of sharing within families while clashing with refugees – demonstrates sharp divisions in notions of home and community in Walker.

Ivanovo, Russia: The Village and the City as Home

Local residents in Ivanovo expressed a strong sense of attachment to their city in general as home, rather than to any particular neighbourhood or community, as in the other two cases. This reflects the fact that industrial ruination and socio-economic deprivation is widespread throughout the city as a whole, rather than clustered in one area. Place attachment to the city was at odds with "outsider" perspectives which tended to see only stigmatization, pollution, and decline, which links with the discussion in chapter 5 about spatialized deprivation and exclusion. For example, I spoke with several Russian people in St Petersburg, where I stayed for an intensive language course before travelling to Ivanovo, to get a sense of what Russian people from outside of Ivanovo knew or thought about the city. People responded with surprise: Ivanovo had a reputation for its dirtiness and ugliness, and it was not well-known inside or outside of Russia as a place to visit for pleasure or tourism.

Roman described the discrepancy between perspectives of people who come from outside of Ivanovo and people, such as himself, who were born there:

> Some people who come from other regions compare the view on our city and they say that it's one of the dirtiest, one of the most bad-looking places

> in Russia. I don't know what it is connected with, but these are their words of people from other regions ...
>
> I rather like this place, as I was born here and maybe the fact that my parents were born in a village and they came to the town to study and work here and it was a great change for them and I appreciate this, and it contributes to my view of life and I feel okay that I live here, although I am afraid of my future as I don't see many opportunities here. (interview, 12 September 2006)

Roman's account reflects tension between insider and outsider perspectives of Ivanovo. He could understand the criticisms of people from other regions, as he was aware of the many socio-economic problems in Ivanovo. However, he was attached to the place connected with his family history. Despite the lack of opportunities he saw in Ivanovo, Roman recognized the achievements in terms of social mobility and quality of life that his family had gained through moving from a village to the city.

Another interviewee, Elena, who had lived in Ivanovo for seven years, also offered a bridge between an "outsider" and "insider" perspective. She admitted that her first impression of Ivanovo was not very positive:

> I have been here for seven years. You know this city is really dirty and ugly. When I came here it was a sort of culture shock. Everything was so ugly. We used to have very many disorganized markets. My impression of the city is getting better. As I have said, the factories are going to be reopened, and it's getting more or less better. (interview, 26 September 2006)

Despite her early impressions of the city, Elena presented a positive perspective on recent changes in the city and on its future. She recounted details of municipal and regional politics and policies, recent newspaper stories, and her knowledge of different textile factories in the city. Despite her awareness of local issues, she was more critical of certain aspects of the city than other residents I spoke with. For example, while other people spoke proudly of Ivanovo's identity as the "Russian Manchester," Elena was more sceptical of the epithet:

> As for the name, the "Russian Manchester," maybe you know that there is an advertising company called "Russian Manchester" here, but sometimes it's just used in an ironic sense, just like a mockery; the way Manchester is

> big and still it has a future and Ivanovo is small and it doesn't have a future. (interview, 26 September 2006)

Elena's insight about people's self-critical and ironic use of the description "Russian Manchester" reveals something about the contradictory nature of place attachment in Ivanovo. Her account also demonstrates a positive city image on the part of residents of Manchester as a Western city that is successful, but without also recognising the fact that Manchester has undergone a difficult post-industrial transformation and that its textile industry is a thing of the past.

Outsider perspectives of Ivanovo often portray the city as a typical depressed and unattractive old industrial city, while semi-outsider perspectives, such as Elena's, provide a bridge towards a more positive view of change and possibility. Admittedly, this image of Ivanovo was what attracted me to the city as a place to study industrial ruination. However, few places see themselves as exemplars of industrial decline: each place has to have its own positive images and associations, a theme which will be further explored in chapter 7.

Many local people had a relatively strong place attachment to Ivanovo. The residents I spoke with said that despite Ivanovo's problems, they loved the city because it was their own. For example, a member of the Veterans' Society described her view of Ivanovo in relation to other cities:

> I often go to other cities of Russia and I often go to Vladimir, which is much cleaner than Ivanovo and more compact, although Vladmir is much larger. It is very difficult to compare Ivanovo with other cities, such as Yaroslavl or Nizhni Novgorod, because those cities are much larger, they have other branches of industry, many more people live there, and salaries there are higher, but nevertheless Ivanovo is my native town and I love it greatly. (interview, 21 September 2006).

Natasha, a local resident with family connections to the textile industry, shared a similar sentiment: "When I compare Ivanovo with other towns of Russia, this comparison does not play in our hands, as other towns and cities are a bit cleaner, a bit brighter and their industries are more developed, and people there are more cultural, but nevertheless I confess that I love this town because I was born here" (interview, 21 September 2006). Natasha also described her pride at living in the "Russian Manchester": "I was born here, I grew up here, and I am very proud of living in the textile city. It was even compared with Manchester and called the 'Russian Manchester' and I've always been proud of this fact."

During one of our driving tours in late September 2006, Andrei spoke about his love of Ivanovo: it was located on a river in the heart of the Russian forest, and it was full of university students. On that day, the sun shone and the autumn leaves were turning bright red and orange. In the brighter light and crisper air, the ruined buildings and pothole-ridden streets seemed less forbidding, and I could see why Andrei felt strongly about his home city. Andrei also reflected on his dismay at finding out that many of the area textile factories were only partially in operation; he had thought that the industry was doing much better. He exclaimed somewhat despondently that Ivanovo was not really the "Russian Manchester" it claimed to be. His interpretation of that nickname was revealing, in a similar way to Elena's comments. He viewed "Manchester" as a positive comparative epithet, one that he seemed to think Ivanovo had failed to live up to. He had a certain image of what it meant for Ivanovo to be "a Manchester," associated with growth, and decline did not fit into that image.

Roman also gave an account of his own contradictory relationship with Ivanovo, a city that he felt attached to, yet one that he knew had few employment opportunities and little security:

> I can call myself a patriot, but I am not the kind of patriot that shouts that he loves his region and his region is the best. It is not the best and it has never been the best. But I believe that something should change or I shall change and I believe in the future things will be better. Ivanovo is my home but I won't hesitate if I get a chance to leave it, to leave this place and go somewhere else just to be sure of my future, just to be sure of my family, of my children, of my future because now I am thinking of it more and more. (interview, 12 September 2006)

Roman's narrative demonstrates the stark contrast between desiring a better future for one's home city and desiring a better future for oneself. This conflict between hopes and expectations, and between image and reality, was a theme in many local accounts.

Place attachment to homes under threat of demolition was also a theme in Ivanovo. In this case, the homes under threat were single-family wooden houses with small gardens, a type which proliferates throughout the region (figure 6.3). These homes are evidence of Ivanovo's pre-twentieth century history as one of many rural villages in Russia with a basis in handicraft and weaving. Many of these houses are in a state of disrepair and rely on owners' individual initiatives for plumbing and fixtures. When I was in Ivanovo, I walked past a cluster of these wooden houses every day get to and from my own accommodation, a Soviet-style block of university flats. I observed residents using a communal water pump in the alley between the houses. In 2000, 64 per cent of dwellings in the Ivanovo region had running water, and only 54 per cent had a hot-water supply (Kouznetsov 2004, 38). In 2006, the wooden houses, particularly those in central areas of the city, were threatened with demolition by developers from Moscow who wished to build more modern residential apartment blocks. According to several of my interviewees (September 2006), residents who remain attached to their homes despite structural problems resisted prospective demolitions. Similarly to residents in Walker, they also suffered from stress and mental anguish at the possibility of losing their homes. But their protests did not amount to the same kind of collective community struggle as in Walker.

Echoing the landscape of industrial ruination, which is spread throughout the city rather than clustered in particular areas or communities, place attachment for residents in Ivanovo was strongly linked to the city as a whole, rather than to particular communities. Place attachment to physical homes was focused on pre-industrial wooden housing that often lacked running water. This shows a stronger tension between the pre-industrial and the industrial, which is still the dominant collective narrative and identity of the city, than between the industrial and the post-industrial, which is still far removed from the socio-economic realities of the city.

Conclusion

In all three case studies, people constructed their homes and communities of the past and present in a way which reflected a sense of the theme "devastation, but also home." In Highland, homes represented a sense of family and community achievement, yet residents were nostalgic for lost social structures of family and work and had experienced long-term spatial, social, and racial exclusion. In Walker, residents regarded their community as something that had endured socio-economic decline and remained strong; they were proud of it and would fight to keep their houses from being demolished. At the same time, local people were aware of the social and economic deprivation within their community. Similarly, in Ivanovo, there was conflict between an idealized image of

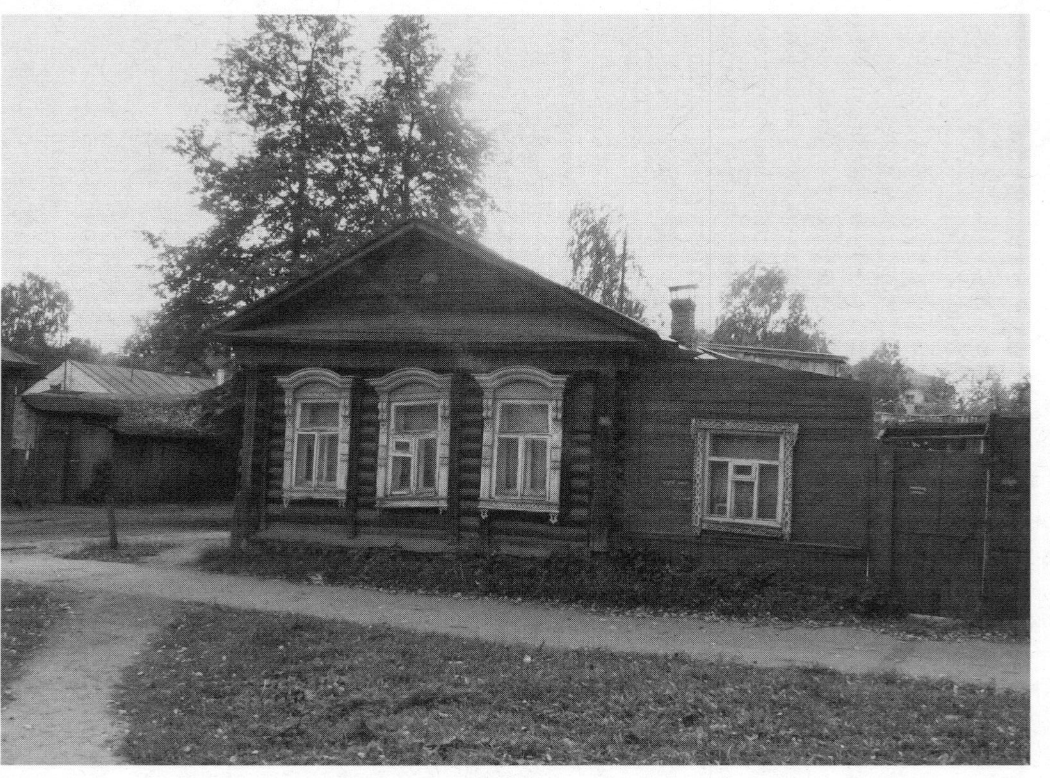

Figure 6.3 Wooden house, Ivanovo, September 2006

the city as a proud and industrious "Russian Manchester" and the socio-economic realities of decline and poverty.

Both Highland and Walker were formerly based on a traditional Fordist model of working-class family life, with men working in the shipyards or chemical factories and women staying at home and raising children. The two communities also had a long history of mutual support between families and neighbours, and adult family members often lived in close proximity to one another. The importance of home for residents in both cases relates to concepts of family, but with the decline of manufacturing work and the fragmentation of the nuclear family in the post-Fordist years, concepts of family and home also represent a form of nostalgia, a way of trying to maintain fading social structures amidst socio-economic change. By contrast, for residents in Ivanovo, the concept of home related more broadly to their notion of the city: as the place in which they were born and raised, worked, lived, and spent their free time. They expressed their attachment to the declining city in conflicted terms, as a result of their intimate knowledge of its social and economic problems. Place attachments to specific homes in Highland and Walker were to houses built during the height of the industrial area, whereas place attachments to houses in Ivanovo were to houses built during the pre-industrial era, which reflects notable differences in the historical legacies of urban and industrial developments.

This research relates to the literature on the psychological impacts of residential mobility in the cases of Walker and Ivanovo. Residents in Walker faced the threat of dislocation, and they suffered from depression, anxiety, and stress as they waited to find out about the future. Similarly, residents in Ivanovo suffered psychologically from the threat of displacement. However, in the case of Highland, residents did not face relocation, but rather faced negative health and socio-economic effects by remaining in their community. Residents continued to build new homes within the area or to improve their existing homes despite the low property value and health risks associated with their proximity to contaminated old chemical factories. This indicates that community cohesion through place attachment is not necessarily a good policy tool for community development, a finding that contrasts with with some of the literature (cf. Brown et al. 2003; Ledwith 2005). In contaminated and impoverished areas such as Highland, which have few immediate prospects for clean-up and renewal, there are significant health and socio-economic costs to remaining in the community. Furthermore, the psychological phenomenon of "adaptive preferences," as Sen (2009) has argued, cannot serve as a justification for neglecting issues of poverty and socio-economic deprivation. At the same time, place attachment was a powerful factor in mobilizing residents in Walker and in Niagara Falls, Ontario, to participate in debates and struggles over local urban planning, which suggests that place attachment can, in some cases, play a positive role in community redevelopment.

Despite the contradictions and conflicts of place attachment in areas of industrial decline, each case also pointed towards positive potentialities of place. These built on particular identities and capabilities distinctive to each place: (1) the ability to work together and support one another in the face of hardship, as in the case of Highland, (2) the ability to resist and shape local politics around redevelopment and change, as in the case of Walker, and (3) the sense of commitment to a city as a home while recognizing its negative features, as in the case of Ivanovo. The diverse perspectives and challenges of people who live through processes of deindustrialization have significant implications for how we might tackle issues of industrial decline, the subject of the next chapter.

Chapter Seven

Imagining Change, Reinventing Place

The act of imagining the nation, city, or community encompasses questions of identity, belonging, aspiration, memory, equality, and social justice. A number of authors have written about the role of the imagination in relation to the city, community, place, and nation, and have explored diverse forms of imagining, re-imagining, and imaginaries (Amin and Thrift 2002; Donald 1999; Gaffikin and Morrissey 1999; LiPuma and Koelble 2005; Shaw and Chase 1989; Westwood and Williams 1997; Zukin et al. 1998). Writers, artists, activists, developers, urban planners, and government officials, among many other social actors, have conceived of different ways to imagine places around the world, and there are examples of such places in urban plans, social memory, art, literature, and film. Dickens' grim portrayal of nineteenth-century working-class London and *Bladerunner*'s dystopian futuristic vision of Los Angeles are classic examples of negative imagined places (Davis 1990; Donald 1999). While imagining place may create exaggerated utopian or dystopian visions, it is also a way to facilitate social change, "an attempt to imagine not only the way we live, but above all the way we live together" (Donald 1999, xi). In *Spaces of Hope* (2000), Harvey makes the case for re-imagining places, emphasizing the importance of utopian thought in the present era, where utopian theories have fallen out of favour. He traces the past failures of utopian and dystopian movements and outlines a new kind of utopian thought, called "dialectical utopianism," which posits a vision of a more equitable world of work and nature that is sensitive to the dialectical dynamics of change. Imagining change and reinventing place are important theoretical and practical acts for people who live within landscapes of industrial ruination and urban decline, as they help people to enter into dialogue with one another and identify challenges and constraints while seeking alternatives and possibilities.

However, the role of imagination in shaping places can have both positive and negative implications, particularly in the context of de-industrialized communities. Popular imaginations of old industrial places are often based on prejudices and stereotypes which have little to do with present lived experiences and social realities. These negative imaginaries can stigmatize places of industrial decline, and contribute to their continued socio-economic isolation and exclusion. As Bourdieu (1999, 123) observes:

> These days, referring to a "problem suburb" or "ghetto" almost automatically brings to mind, not "realities" – largely unknown in any case to the people who rush to talk about them – but phantasms, which feed on emotional experiences stimulated by more or less uncontrolled words and images, such as those conveyed in the tabloids and by political propaganda or rumor.

Similarly, Wacquant (1999, 131) argues that catch-all notions of working-class neighbourhoods in decline (in the United States) such as "neighbourhood," "housing ghetto," and "immigrant ghetto" represent a "largely fantastical discourse" which contributes to "the spiral of stigmatization that leads to branding working-class housing projects as condemned places synonymous with social indignity and civic relegation, thereby adding to the burden of symbolic domination that the residents of these housing projects must bear *on top of* socio-economic exclusion."

Indeed, the public imaginaries and representations of places in decline only compound existing problems of socio-economic deprivation and exclusion through destructive stigmatization. For example, a recent study (Brent 2009) of a deprived estate in Bristol in the United Kingdom explores the destructive nature of outsider representations of deprived communities by drawing on academic arguments about community, deprivation, and youth work, as well as the author's personal experience as a youth worker in the estate for almost thirty years. The book includes critical reflections on contested ideas about community, and on the tensions between insider and outsider accounts of communities. Brent explores the impact of negative official and media representations of deprived communities from outsider perspectives, as contrasted with insider perspectives from community plays and radio, and his own role as an "outsider within" as a public servant in the area. The research is critical of short-term solutions to youth "problems" which are the product of long-term social problems such as inadequate housing and a lack of amenities.

In another study that reveals the importance of positive versus negative city images, Zukin et al. (1998) use the concept of the urban imaginary to analyse how the decline of Coney Island and the growth of Las Vegas as public spaces of amusement relate to their different urban imaginaries, respectively as low-class and high-class spaces, and are linked to different eras of capitalism and to different forms of racialization. The authors define the urban imaginary as "a set of meanings about cities that arises in a specific time and cultural space: first, the relational – and often hierarchical – meanings places hold in the popular imagination, and second, the connections between hierarchies of place meanings and hierarchies of social class and race" (Zukin et al. 1998, 629–30). My analysis follows Brent in criticising the destructive power of outsider perspectives on areas of decline and deprivation, and follows Zukin et al. in exploring connections between social and cultural imaginaries of place with real social hierarchies of class and race, and with tangible impacts on socio-economic growth and decline. Given the role of urban imaginaries in shaping urban realities, it is important to find alternative, positive ways of imagining places and futures, particularly in areas marked by socio-economic deprivation, which often suffer from stigmatization.

As this chapter will show, there are different ways of imagining change and reinventing place in the narratives of people who dwell in landscapes of industrial ruination and urban decline. Much of the literature on regeneration and urban planning adopts the idea of the imaginary through focusing on developing policy visions for future cities (cf. Cochrane 2006; Morrison 2007; Newcastle City Council 2008; Tallon 2010). However, there are many underlying assumptions in these regeneration visions, including, for example: that they exist in a context of economic growth rather than decline (especially before the 2008 recession); that the model of arts- and property-led regeneration is a one-sizes-fits-all solution to socio-economic decline; and that the post-industrial economy based on knowledge, services, and information is the best ultimate future for any city in the developed world.

In all three cases I have explored in this book, people and places suffered from the stigma attached to industrial decline: social and economic deprivation in Walker, Soviet and industrial ruins in Ivanovo, and toxic contamination in Niagara Falls. Individual narratives, many of which came from people living outside of places affected by crime, drugs, poverty, and other negative features, reinforce this stigma. Symbols, such as Love Canal, Swan Hunter, or Soviet statues, solidified these negative images, even as they also carried more complex meanings. In each case, people attached symbolic – rather than economic – value to certain sites. Love Canal exemplified the devastating impacts of industrial contamination, Swan Hunter, as the last shipyard of the Tyne, symbolized the pride of shipbuilding, and textile factories in Ivanovo embodied the heart of the "Russian Manchester." The symbols themselves are also revealing. Love Canal represents the devastating impacts of industrial decline and toxic contamination that have occurred throughout the region of Niagara Falls. The stigma that Love Canal brought to Niagara Falls has silenced the issue of contamination in the region, and so the toxic legacies of multiple "Love Canals" have largely remained unexplored. The epithet of the "last shipyard of the Tyne" conveys an expectation of future loss, of holding onto something noble and proud that stands alone. The embodiment of "Russian Manchester" through a vast number of abandoned and semi-abandoned textile factories illustrates the tenacity of industrial identity despite economic decline.

Despite the contradictions in place imaginings of past and present, each case also points towards positive potentials of place. These are built on particular identities and capabilities distinctive to each place: (1) the ability to resist and shape local politics around redevelopment and change, as in cases of Walker, Newcastle, and Niagara Falls, Ontario; (2) the ability to maintain particular identities (economic and social) despite pressures to change, in creative and pragmatic ways, as in Ivanovo; and (3) the ability to work in partnership and to share knowledge, as in the case of community volunteer work with asylum seekers in Walker and grass-roots community projects in Highland.

The local politics of community and development provides clues about the legacies of industrial ruination: how people manage their current situations, what they value and will fight for, and how they relate to processes of change. In each of the case studies of industrial ruination that I explored – Niagara Falls, Canada/USA, Newcastle upon Tyne, United Kingdom, and Ivanovo, Russia – ways of imagining change and reinventing place were embedded in local politics, with tensions between capital, community, and the state. Two opposing yet interrelated themes emerged across the case studies: the politics of contestation and the politics of resignation. Both themes relate to the different capabilities of local residents for action, based on the particular political legacies associated with processes of industrial decline. Through highlighting specific stories of the local politics of community and development in each case, this chapter explores how people imagine processes of socio-economic change in landscapes of industrial ruination and how they re-invent notions and realities of place. It also reflects on the implications of imagining place for policy actors who are interested in tackling challenges and issues associated with deindustrialization and redevelopment.

Niagara Falls

The politics of community and development surrounding abandoned industrial sites and adjacent communities in Niagara Falls revolve around competing visions from municipal government, corporations, community organizations, local residents, trade unions, and activists. The unknown element of contamination and the limited capacity of governments, scientists, or concerned citizens to research, understand, fully remediate, and integrate contaminated areas have complicated many of the debates and struggles. There is also tension between creating economic growth and fostering a clean environment. As I discussed in chapter 2, the debates surrounding the proposed development of a skating arena on a portion of the former Cyanamid property are an example of these contested politics. The more tenuous brownfield initiatives in Highland Avenue also illustrate the constraints on local political action in areas of extreme marginalization.

In Niagara Falls, Ontario, many people imagined change in the city through the lens of the "post-industrial" economy of hotels, casinos, and tourism, and through their faith that brownfield redevelopment would be a safe and healthy way to move forward from a toxic industrial past. Other people could not imagine change without also remembering the pain of past injustices. The contested politics of imagining renewal and change in Glenview-Silvertown, Niagara Falls, are perhaps most vividly illustrated through the case of the proposed arena development on the former Cyanamid site, which was discussed in chapter 2.

The trade union Unite Here, which represents hotel, casino, and other service sector workers, took up the concerned residents' case against the building of the arena. In June 2006, it issued a press release that highlighted concerns about the lack of tests by people other than those hired by Cytec, and the low level of actual clean-up of the site. As a result of the intervention of Unite Here, Niagara Falls City Council held a number of meetings between 2006 and 2007 to try to address these concerns. A culmination of these local events occurred at the special council meeting regarding the proposed arena development on Cytec property on 11 April 2007, which I attended as part of my research. The core of the meeting consisted of a series of presentations based on a new report written issued only the night before, entitled, "Response to Issues Raised by Unite Here, Proposed Arena Complex Site." The presenters were Cytec consultants, experts in science and in economics, and they produced scientific and economic data with graphs and figures on slides to show that they had gone through the correct planning processes, had done enough testing, and had found no concrete evidence that a risk to human health would be associated with building an arena on the proposed site. Further support came from community members representing hockey and skating associations, who expressed their desire to have a community arena and their disapproval of Unite Here's opposition.

The opposition itself was armed with few members, fewer economic resources, and no scientific experts. Their case rested solely on accounts from residents such as Jim and on legal and political support from Unite Here. While the lawyer representing Unite Here was supposed to give a critical presentation, he refused in light of "new information" revealed in the city council's report. This refusal provoked debate and resistance from the mayor, the clerk, and some of the councillors, none of whom were supportive of Unite Here's position. The only previously supportive city councillor commented that her eyes had been opened: she felt that if residents could see a broadcast of the meeting, their misconceptions and fears surrounding the proposed arena development would be resolved. The mood after the meeting was one of clear victory for Cytec and the city council and defeat for Unite Here.

Throughout the meeting, different parties grappled to gain the upper hand through appealing to scientific, economic, and legal discourses. The city council and Cytec responded to the concerns raised by Unite Here with a spectacular arsenal of scientific and economic expertise. I noted some flaws in the "expertise" presented: there was speculative economic accounting for job creation, and test figures only related to the sixteen-acre arena site and not to the whole ninety-three-acre site. However, their presentation made it clear that it was difficult if not impossible to contest their scientific claims without extensive financial and political resources. I interviewed the lawyer for Unite Here later that evening (11 April 2007), and he explained that he had refused to speak because the city council had introduced a surprise attack when they released the new report. He explained that he was setting the grounds for an appeal on the basis of natural justice, as it was unfair for the city to present a new basis for discussion without prior agreement. Nothing seemed to come of this claim. On 8 May 2007, an article in the *Niagara Falls Review* noted that the city council had voted unanimously to build a four-pad arena complex, "beefing up the original plan for the twin-pad on the arena site" (Larocque 2007). Residents' accounts regarding the health effects of Cytec/Cyanamid during its days of operation were compelling; however, it was impossible to assess the health impacts of lingering contamination.

The concerned residents and trade union representatives who waged battle with the city council over their brownfield redevelopment arena proposal were soundly defeated at the council meeting. Some of the residents set their concerns aside in the face of new evidence, presented by Cytec and the city council, about the safety of the site. But others such as Jim, who had dedicated decades to campaigning against Cytec's corporate negligence of toxic pollution, were not about to give up the fight. These conflicting imaginaries of local change highlight the power of place memory and the psychological effects of contamination on people. In some sense, fighting a battle in the present after suffering in the past is a way to ask for compensation for past injuries, but without knowing in what terms to phrase them and what battles to choose. In this respect, the battle over the arena was as much symbolic as it was material.

In contrast, in Highland, Niagara Falls, New York, the community engaged in small-scale grassroots efforts to reinvent itself. For example, Carl, the manager of the Highland Community Revitalization Committee, spoke at length about some of the accomplishments his organization has achieved, including: partnership with Niagara Falls City Council in the Highland Avenue brownfield plans; involvement with a community outreach program run by Niagara University; and greater community attendance at city council meetings (interview, 23 March 2006). He was also excited about the prospect of a National Heritage designation in Highland Avenue which would draw on the history of the area as part of the "fight for freedom," in which people in the North helped enslaved black people from the South escape via the Underground Railroad into Canada before the American Civil War.

However, others were more sceptical about the possibilities of change in their community. When I asked one Highland resident about how she felt regarding the city's plans, she seemed unaware of them: "I really haven't been reading the paper about that ... But whatever they plan, it takes them too long to do" (interview, 28 March 2007). Another resident was cynical of the prospect of change, and referred to the initial redevelopment plans from close to ten years earlier:

> You're looking at your chart of the Highland Redevelopment Plan, which is just somebody's idea of buying votes, they never intended for this thing to happen anyway. My first knowledge of the plan was when I opened up my mailbox and looked at the flyer that they had that showed that they had redirected the street that I lived on and that my house was gone [*laughs*], and I was rather upset at first. How dare these people move my house and not even tell me what their plan was? But then when I finished reading it and I saw who was putting this thing together, I laughed, I said, I don't have to do anything; this'll never happen. It's just politics; it's just poor politics. (interview, 21 March 2007)

This resident voiced two criticisms: first, of the projected change that would destroy his house, and second, of the impossibility of real change in the context of extreme community poverty and ineffectual local government. Thus, despite small steps towards imagining change within the Highland community, there was a sense of scepticism and resignation in the face of the grave and enduring problems of contamination, stigmatization, and socio-economic deprivation.

The only visible signs of struggle that I found in the Highland community were the protests of the Laborers' International Union of North America Local 91, a Niagara Falls-based trade union that represents labourers in construction, brick masonry, asbestos removal, hazardous materials abatement, paving roads and runways, and pipe installation. Reports in local newspapers described the actions of the trade union as mafia-like and cited incidents of violence and intimidation against its enemies, including reporters and government officials (Moretti 2007). In spring 2007, the trade union demonstrated outside a site where new public housing was about to be constructed, to protest hiring non-unionized, non-local labour for the project. Media reported that the trade union had a "vice grip on construction jobs and development" (Moretti 2007). Given the context of high unemployment and socio-economic deprivation in Niagara Falls, New York, it is not surprising that one of the few contentious issues in Highland hinged on the only major employment opportunity in the area. Since the time of my research, there have been some developments within the Highland

Community Brownfield Opportunity Area Program, including a few open house community engagement events and the creation of a resource centre (The Brownfield Opportunity Area Program 2008), but there has been little evidence of real change.

The case of Niagara Falls as a whole vividly demonstrates the gravity and neglect of toxic landscapes and legacies of industrial ruination. This has implications not only for Niagara Falls but for contaminated areas around the world, including those in the North American Rust Belt, industrial China, and other cities with heavily polluting industrial pasts and presents. In Niagara Falls, the most critical step in addressing this issue is to facilitate the clean-up of contaminated sites by any means possible: by combining government, city, and community resources; by seeking help from environmental agencies and actors; and by drawing public attention to the enduring toxic legacies of contamination rather than to developing the tourist industry.

If Niagara Falls is to resolve the issues surrounding toxic legacies of industrial decline on both sides of the border, it must address two issues: accountability and recognition. Accountability was the greatest factor influencing corporate and governmental claims regarding the health impacts of the contaminated sites in Niagara Falls. In both cities, most residents failed to hold anyone legally accountable for their health problems, as a result of insufficient efforts or legal defeat. Hooker Chemicals was held legally accountable for Love Canal, and Norton Abrasives was brought to justice over dumping in the Welland River. However, these flagship cases involved tremendous local political struggles, and the vast majority of contamination cases are unresolved.

Residents of both communities were also battling for recognition – through heated city council politics in the Cytec/Cyanamid case, and through voicing tales of injustice in Highland. Amidst the upheaval of their lives and communities, residents wanted their stories to be listened to and their difficulties acknowledged. At the local level, a minimal gesture of public recognition could take the form of a monument or plaque to mark the sites of Love Canal and the Chippawa blob as spaces of trauma, or that of a dedicated museum, with possible funding from national and international environmental agencies. On a wider level, recognition could take the form of a public apology from corporations and governments to those who have suffered from the legacies of toxic contamination. However, while accountability is a practical impossibility in most cases (as the financial cost of contamination is difficult to quantify, and there are many cases), recognition is a minimal gesture from political authorities. To address the sense of injustice felt by many people over the myriad health and socio-economic effects of toxic legacies in Niagara Falls, a gesture beyond recognition, but perhaps short of full economic accountability, might be a positive way forward. This might constitute a small out-of-court financial package to residents and former workers from corporations or local governments based on recognition of past injustices. Through public recognition that goes beyond mere gestures, the deep injuries of toxic legacies might begin to heal. However, the wider question of how to reverse economic decline and high unemployment also needs to be addressed in order for people in Niagara Falls to truly move "beyond the ruins" (Cowie and Heathcott 2003).

Walker, Newcastle upon Tyne

The East End of Newcastle, and Walker in particular, has a strong tradition of local political engagement (Madanipour and Bevan 1999, 30). The Labour councillors of the ward have maintained close ties with the area and local residents. Decline in the manufacturing industries along the industrial riverside and in the East End as a whole has meant a shift in the type and nature of jobs towards piecemeal, flexible, and largely non-unionized labour. Thus, community residents, rather than workers on the factory floor, have been the main actors in local struggles over processes of social and economic change. The local politics surrounding community regeneration in Walker were still tied to the history of labour politics, but the worker base was eroded, and the community spirit of contestation and solidarity was all that remained of the industrial workers' past. Most Walker residents primarily related to the industrial legacy of shipbuilding through the prospect of imminent regeneration. They were concerned with the implications of regeneration for their community, which was based on a shared industrial history that had been in perpetual decline for over forty years. Thus, in 2006, the politics of contestation among residents was closely connected with the concept of community cohesion as a legacy of industrial ruination.

The local politics of post-industrial change in Walker have been fraught and contested. Newcastle City Council's role in regeneration and economic development processes was clearly evident. At the same time, other actors such as faith communities emerged in the absence of other leadership. For example, the Anglican Church has developed the Urban Ministry Theology Project, which provides its own analysis of the issues of social and economic exclusion and deprivation in the East End and has engaged with issues of homelessness, poverty, addiction, and gambling through various church projects. The Roman Catholic Church has also been active in the community, and has worked with the school and provided skills training in numeracy, literacy, and IT (with the support of government funds in the 1990s), care for the elderly and needy, and support for working families. The leadership role of faith communities in Walker's social and economic welfare has come about through the absence of other actors, such as employers and the state, which customarily fill such roles in the context of modern capitalism. Although the church has historically played a role in local community welfare, it has done so alongside other actors. In Walker, the increased role of churches reflects the diminished ability of capitalism and the welfare state to deal with social and economic problems associated with industrial decline, at least during this critical time of instability and transition.

During my research, I observed that the power, voice, and authority to shape the regeneration of Walker Riverside extended from the government and developers at the top to the community and voluntary sector at the bottom. In other words, the government and developers spear-headed the proposed regeneration plan and engaged with the community and other partners at further stages in the approval process. There was only relatively extensive community consultation because of resistance from the local community to proposed regeneration during the early stages. Some community and voluntary sector organizations have begun to foster community development strategies "from below," but these efforts have remained tied to the funding and institutions of the city council and its development partners. Local politics in Walker have been post-industrial in the virtual absence of a labour or trade union politics on the factory floor. However, the resilience of the local community in its struggle to retain its homes, families, and sense of community represents a legacy of the industrial past, based on the spirit of solidarity and resistance of workers and workers' families.

As discussed in previous chapters, Newcastle City Council-led regeneration went ahead in 2006 after a long series of heated debates about the proposed demolitions. The council targeted Walker for regeneration because it was a deprived area, stigmatized as a place with "dependency culture," low aspirations, social problems, and high rates of unemployment, crime, drug and alcohol abuse, and people on benefits (MacDonald 2005). Stigmatization was an important issue in the regeneration debate, for while outsiders perceived the area as a hot-bed of social ills related to socio-economic deprivation, the community itself, while recognizing socio-economic problems, had a strong sense of solidarity and collective pride, particularly in its homes (as discussed in chapter 6). After a long period of community consultations, residents were asked to choose one of three options: minor, moderate, or major impact: the minor option involved no housing demolitions, the moderate involved several, and the major involved a considerable number. The third option was the city council's clear preference, and the only option under which the city's regeneration partner promised the community investment in schools, local infrastructure, and shops. The idea was effectively to plan regeneration – and gentrification – in the area through property development. As Smith (2002, 427) observes: "The process of gentrification, which initially emerged as a sporadic, quaint, and local anomaly in the housing markets of some *command-centre* cities, is now thoroughly generalized as an urban strategy that takes over from liberal urban policy." While early forms of gentrification between the 1960s and 1980s were relatively sporadic and organic, and emerged from various urban forces, such as inner city decline and artists' loft-living (Zukin 1991), by the 1990s gentrification had become entrenched and generalized as a "crucial urban strategy for city governments in consort with private capital in cities around the world" (Smith 2002, 441). However, the city council and its regeneration partner underestimated the endurance of negative place images: the middle classes in Newcastle upon Tyne were not suddenly convinced that Walker was a new "place of choice" (as the regeneration lingo put it) because of new housing, especially as the development of services and other infrastructure was dependent on housing sales.

In July 2009, three years after "regeneration" began in Walker, all of the old houses had been demolished and many of the new houses had been built, but almost none had sold. Follow-up interviews with a Labour city councillor, a city council officer for economic development, a former city council regeneration manager, and two local residents (8–9 July 2009) revealed that the plan was a failure. There were critical housing shortages in the area, and increasing racial tension between incoming asylum seekers with refugee status and the "traditional" predominantly white, working-class population. The financial crisis that began in 2008 was one of the reasons cited for this failure. The United Kingdom Audit Commission released an independent report on neighbourhood regeneration in Newcastle in May 2009 which criticized the city council for the fact that no new shops and limited community facilities had been built in Walker, and for "significant gaps between the most deprived and least deprived areas in the City" (Audit Commission 2009). Then in March 2011, the regeneration funding came to an abrupt end. The United Kingdom coalition government cut the Housing Market Renewal Pathfinders (£2.75 billion between 2002 and 2011) program that funded regeneration six years earlier than had been expected in the spending review of October 2010 (Long and Wilson 2011).

These developments demonstrate how drastically redevelopment policies affect people's lives, often in unintended and undesirable ways. They also highlight the failures of one-size-fits-all regeneration strategies that rely on arts- or property-led development to lead the way for social and economic development, rather than prioritizing people. But much of the blame for the failures of regeneration was placed on the context of economic recession rather than on short-term urban policy visions. Economic recession is clearly an important context in which to rethink regeneration strategies, but it is not the only factor: area-based housing-led regeneration has failed repeatedly in the West and East Ends of Newcastle since the 1980s.

The vocabulary surrounding the regeneration process changed to reflect the shifting expectations of the planners. Initially, the Walker regeneration plan was called Community Focus (2005/6), but in 2007 it was renamed Heart of Walker, representing a harmonious and idealized vision of a regenerated community. Unlike the previous campaign, which focused on justifying the regeneration process on the basis of significant decline in the area, the new slogan highlighted a new local vision:

> Walker is a great place to live. It is not just about housing, it is a lifestyle. Only minutes from Newcastle city centre and the quayside it is certainly a great place to be. With new homes, shops, schools and a host of other facilities on the way, it is a wonderful community and a first class location. (Newcastle City Council 2007)

This statement presents the regenerated landscape of Walker as though it already exists. In 2009, the Heart of Walker campaign became "passion-pride-potential." The new slogan is more cautious, and reflects the failures of the regeneration project. The website refers to ongoing improvements and targets the local population as well as newcomers: "The Walker Riverside is being improved to become a location of choice. Current residents will choose to stay in the area, people looking to move into the east end of Newcastle will want to choose to move to Walker Riverside" (Newcastle City Council 2009). The 2008 regeneration plan for Newcastle in 2021 explicitly highlights the importance of image in regenerating cities: "Image matters – negative images of people and places can seriously undermine regeneration efforts" (Newcastle City Council 2008, 13). This is a lesson about failed regeneration efforts in Walker, where negative associations persist despite efforts to promote positive alternative visions. Reflecting on the city council's efforts to re-brand the area as a "place of choice" to outsiders, the Labour councillor for Walker commented about the importance of image for community regeneration:

> Walker has always been a place of choice for people who live in Walker or were born in Walker. People don't want to leave. Bringing people in from outside is a completely different issue because it has a reputation of being, well it is, the most deprived ward in the city. But it also has a reputation for having, rightly or wrongly, high crime, poor health, antisocial behaviour, litter, dirt, drunken youths, rampaging dogs, whatever, you know, everything you can associate with a so-called sink estate, you would get that view from outside. (interview, 8 July 2009)

The Walker case offers several important policy insights for city and regional urban policy actors. First, it reveals that the "post-industrial" strategy of housing-led community regeneration has limited potential for economic and community development. The issues of employment opportunities, skills retention, and skills training within the local community need to be addressed in any plan that would impact redevelopment. Second, it is important to create feasible and sustainable regeneration plans that involve developing not only property but also people. The prolonged nature both of industrial decline and of unsuccessful regeneration projects has made local people sceptical of and resistant to change. Third, recognizing local place identity during processes of socio-economic transition is very important. For example, a museum of shipbuilding designed with community input or a locally constructed monument to shipbuilding would represent community-based commemorative collective action, which could ease a transition by fostering a sense of local heritage based on commemoration rather than loss.

Finally, policy makers could identify and develop ways in which place "works" well in Walker in order to build on existing community assets. Examples include not only the abstract and problematic notion of community solidarity, but also: organizational ability to resist and shape local politics around redevelopment; the six-year battle over regeneration plans demonstrated the resolve of the community around issues that were "close to home"; and willingness to adapt to new circumstances and to share knowledge, evident in the local community's ability to share resources and networks in times of hardship. A positive concrete example of place working well is a former city council regeneration manager's creation of Building Futures East, an independent innovative skills training centre on the riverside, in May 2007. The centre runs free social enterprise training programmes in construction, horticulture, and social care for people of all ages, but focuses mainly on people between the ages of sixteen and twenty-five who are unemployed and typically on benefits. The centre plans to expand into a social enterprise and to train people for prospective jobs at a new renewable wind energy riverside facility. The philosophy behind this project is fundamentally different than that of the city council, with a focus on developing people rather than property, and on creating local jobs and developing local skills (interviews, former city council regeneration manager, August 2005 and July 2009).

Ivanovo, Russia

The new and powerful role of capital in shaping economic realities and the related role of government in guiding and directing investments characterize the local politics of capital and community in Ivanovo. The former role of the state disappeared with the collapse of the Soviet Union, along with its role in the local provision of services and socioeconomic health. Many people have been slow to move away, both socially and economically, from the industrial past. Cultural, social, and spatial traces of the Soviet past also remain in the present, and perhaps this accounts for the passivity of local people in relation to imagining change. The most active political voices have been conservative, from pensioners and nationalist youths who lament the problems of alcoholism, drugs, broken families, and gambling which they associate with the new capitalism and Western influences.

A former textile worker could not remember any strikes during her years as a worker, but she remembered people trying to change their lives through less confrontational means, such as petitions and meetings:

> I can't recollect anything like a strike or a struggle. The only thing we had were meetings when we gathered together with workers from different

> factories and plants. We discussed our problems and nothing else followed, so nothing changed. We were not happy and the workers are not happy now, but we couldn't do anything. The only thing we could do was to gather and to discuss our problems. We had some leaders. We tried to change some things in our lives, we made petitions and papers with our requirements and our desired changes, but they were neglected and we could not change anything although we never tried to fight with our employers. The role of the workers and the residents in guiding their own lives was minimal, and it practically remained the same. (interview, former worker at NIM factory, 12 September 2006)

This narrative contrasts with the themes of indifference and a "Russian mentality" to change, in that the interviewee emphasizes not only people's desire to make changes, but also their inability to create them, because what demands they made "were neglected." However, this account also suggests that people did not and would not try very hard to make changes: "We could not change anything although we never tried to fight our employers." The idea that things "practically remained the same" suggests a sense of continuity with the Soviet past, in terms of local politics. Thus, in this case, the local politics of weak political voice also represents a legacy of industrial ruination.

Another resident, Anna, connected the lack of local political engagement with a culture of indifference and complacency, which she felt the Soviet years fostered:

> There have been no strikes or struggles connected with the fact that people didn't like the political or economic points of their life, but this depends on our people's inactiveness and indifference. For the years of Soviet ideology, all Soviet people were taught to obey the rules and not to think for themselves. That's why people from my generation especially don't know how to hold their own point of view, don't know how to protect themselves. (interview, 21 September 2006)

It is interesting that Anna argued that people from her generation "don't know how to hold their own point of view," yet she did not explicitly position herself outside of this indifference. Nonetheless, despite references to continuity with the Soviet past in terms of mentality and culture, Anna was critical of the political and social restrictions of the period. In recounting this theme of inactiveness, indifference, or inability to create change, this narrative seems to become a self-fulfilling prophesy.

Some of the most active political voices in Ivanovo in recent years have been pensioners. One of the only effective political protests during the post-socialist transition occurred when pensioners took to the streets to demand payment of their pensions. A member of the Veterans' Society described this victory:

> Some time ago, just after the *perestroika* when the economics of our country was in very bad shape, the pensioners didn't get their pension in time. So that's why they came out on the road, just blocked the traffic. This was the greatest meeting that they had. They were effective and all the money was paid in time. And now the pensions are being increased. Not great, but it is slowly increasing. (interview, 21 September 2006)

The Veterans presented one of the strongest narratives of change; they had definite views, and people seemed to respect their authority as *babushkas*, although this respect for the elderly was continuous across both Soviet and post-Soviet eras.

Another of the most active local groups in campaigning for economic and political change in Ivanovo today is the nationalist, pro-Putin youth movement, Nashi (*nashi* is translated as "ours," but it also has connotations of "motherland" and "nation"). However, Nashi is a national youth organization, not a locally inspired group, and has militarist, nationalistic, and conservative roots. Its nationalism and conservatism is an interesting development in post-Soviet Russia (see Malysheva 2007; Schwirtz 2007). Throughout 2006, a small group of students and young people under the Nashi banner campaigned in the streets over several pressing issues in Ivanovo. One protest, called "we want three," drew attention to the economic impossibility of raising a family with more than one child (9 October 2006). Another protest targeted alcohol companies for the debilitating effects of widespread alcoholism on youth, education, and families (13 September 2006). I witnessed one of these protests outside of a casino on Prospect Lenina on 22 September 2006, where a group of about twenty-five young people protested in the street with placards and Russian flags against the growing presence of casinos and their negative effects on family, national values, and the local economy.[1]

The post-industrial strategy of arts- and culture-led regeneration and growth in knowledge, information, and services has clearly failed to alter the fortunes of many old industrial cities, and local economists in Ivanovo have explicitly rejected the Western post-industrial model that holds that society should develop through pre-industrial, industrial, and post-industrial stages, as "inapplicable in the case of Ivanovo as industry should first be restored" (interview with two local economists, Ivanovo regional government, 26 September 2006). They emphasized that the only role that government could play in economic development was in marketing the city as a place for investment: business investment alone could decide the economic future of Ivanovo. Their marketing strategy was to play on the fact that Ivanovo "is not so far from Moscow," where they said 85 per cent of the country's wealth is located. They suggested that Ivanovo could attract business investment in industries and administrative centres, whereas other towns and villages in the region, such as the nearby picturesque village of Plyos, could be marketed for leisure and tourism. Indeed, the regional administration in Ivanovo seemed to have adopted at least one strategy of the competitive post-industrial city: to vie for capital resources through place-marketing. The marketing strategy has a long way to go, however, given that the assistant to the regional governor did not speak English, the websites for the city are predominantly in Russian, and investment in the region is primarily from Moscow.

Drawing comparisons with other old industrial cities also offers valuable insights for Ivanovo. The fact that many people I spoke with seemed unaware that Manchester had also undergone a period of severe industrial decline and subsequent (albeit uneven) regeneration spoke to a lack of knowledge about processes of deindustrialization in other cities. "Russian Manchester" was a positive comparative epithet for Ivanovo, negative only in the sense that it might be failing to live up to the Manchester ideal. In particular, city planners and officials would benefit from a comparison with recent developments in Łódź, the "Polish Manchester." Łódź has only just started to emerge from serious decline, and this provides an interesting point of contrast with Ivanovo.[2] The economic situation in Łódź has remained difficult despite attempts to place-market the city and to adapt to the global economy (Kaczmarek and Young 1999; Liszewski et al. 1997). However, since 2006, marketers have branded Łódź as an emergent city of transformation, largely due to its status as the second largest city in Poland and the recent introduction of cheap international flights. At the May 2007 Global City Conference in Lyon, France, Łódź was described as a triumph of place marketing, signified by the opening of a redeveloped factory site (Morrison 2007). Finally, positive imaginings of place in Ivanovo also provide clues about its assets, such as the creativity, pragmatism, and place attachment of people in adapting to processes of change; in transforming textile factories into new uses; and in persevering with work and life despite considerable socio-economic hardships – all of which are qualities which could be built on in strategies for urban renewal.

Conclusion

The role of the imagination is crucial for shaping urban development trajectories. Ways of imagining change, futures, and cities directly affect people's lives in both positive and negative ways. In all three cases, people and places suffered from the stigma attached to industrial decline: social and economic deprivation in Walker, Soviet and industrial ruins in Ivanovo, and toxic contamination in Niagara Falls. The negative perceptions and imaginaries of people who live outside of these places reinforce this stigma. There is also a moral dimension to this stigmatization, based on the notion of individual responsibility for one's welfare (rather than structural factors) prevalent in "welfare-to-work" discourses about unemployment, homelessness, and social dependency (Carpenter et al. 2007; Peck 2001). One of the main challenges in imagining change and reinventing place in landscapes of industrial ruination is to overcome negative place images and prejudices.

Places of industrial ruination also suffer from a different type of preconceived imaginary: the post-industrial promise of the knowledge and information economy. The uniform application of post-industrial models fails to consider local specificities and tends to privilege physical development over socio-economic and cultural development. This was clearly the case in Newcastle City Council's vision of regeneration as a form of planned gentrification in Walker. By contrast, local policymakers in Ivanovo evaluated the post-industrial model as an option, yet deemed it inappropriate in the context of the city's industrial heritage and socio-economic features. Similarly, the socio-economic and health costs of contamination were a considerable barrier to following a viable post-industrial path in Niagara Falls.

These case studies illustrate the limits of the post-industrial imagination as an immutable model for moving beyond the ruins. Arts- and property-led regeneration has helped to develop many struggling old industrial cities and communities. Similarly, the service sector and knowledge economy have provided a prosperous path for urban development for many other cities. However, these forms of regeneration have limited potential for economic and community development. Physical renewal is not enough; social and economic issues of employment opportunities, skills retention, and skills training within the local community need to be addressed in any plans which would have real impact for urban development. In landscapes of industrial ruination, the implications of arts- and property-led development are not necessarily the same as in contexts of economic growth, even during times of general economic prosperity, for, as Smith and Harvey (2008) argue, capitalism produces inherently uneven geographies. Capitalist urban development is premised on profit, and implies "creative destruction," demolishing buildings and displacing people to make way for new people and places. But the "inevitable" destruction from capitalist development is not necessarily negative in the long-term, as it creates the possibility of clean, functioning, and healthy cities. For example, the mass clearances of insanitary slum dwellings in the development of European cities in the late nineteenth-century displaced many working-class populations, yet paved the way for public transport, gas and electricity, roads, and modern housing (Briggs 1968). Nonetheless, it is important to recognize that while the costs of displacement may be similar in different cases of property-led development, the benefits may not. In the highly precarious and uncertain contexts of industrial ruination, developers can simply abandon urban renewal projects when things go wrong, as in the case of the 2006 planned regeneration of Walker.

In landscapes of industrial ruination, there is no single imaginary of change that will produce prosperous, liveable, and equitable communities. Rather, imaginaries are diverse and contradictory, situated in particular times and places, and linked to different social actors and interests. Imagining change and reinventing place in both creative and constructive ways is an important challenge facing both policy actors and the people who inhabit landscapes of industrial ruination.

Chapter Eight

Conclusion

As symbolic former industries, Cyanamid and Union Carbide in Niagara Falls, Swan Hunter in Newcastle upon Tyne, and the Big Ivanovo Factory in Ivanovo encapsulate the struggles, failures, and hopes of bygone industrial eras. Their stories reflect core themes in industrial ruination and urban decline. The enduring legacies of the Highland Avenue brownfields and the Cyanamid dust are reminders of the devastating impacts of industrial decline and toxic contamination throughout the twin-city region of Niagara Falls. They represent multiple, hidden "Love Canals" submerged beneath the casinos, hotels, amusements, and the great natural wonder of the falls. By contrast, Swan Hunter represents nobility and pride as the "last shipyard of the Tyne," yet its epithet carries an expectation of inevitable loss. Lifelines kept the last shipyard afloat for as long as possible, and its final closure in 2006 signalled the official end of an era. Finally, the Big Ivanovo Factory, a flagship textile factory that is partially abandoned and partially in operation despite economic setbacks, highlights the gaps between expectations, hopes, and realities in a depressed post-Soviet industrial city that still holds onto its historical industrial and Soviet identities. Each of these sites speaks to the trauma, uncertainty, and tenacity of lived experiences with painful post-industrial transformations.

On a first reading, landscapes and legacies of industrial ruination and urban decline suggest devastation, turmoil, waste, destruction, abandonment, deprivation, contamination, and neglect. But these negative associations are only part of the picture. Much of the characterization of places as landscapes of devastation is reinforced by negative perspectives and stigmatization from outsiders to the deindustrialized communities. In each case study in this research, local people related to devastated landscapes with strong place attachment and positive place images. Through the lens of "industrial ruination as a lived process," this book has revealed life-affirming themes amidst the devastation: collective and individual strategies for coping with painful transitions in work and life; strong local networks of support and solidarity within placed-based communities; and ways of imagining renewal and change that challenge prevailing models of post-industrial change.

Different theoretical approaches to industrial ruins, of urban planners, political economists, photographers, and dereliction tourists share a common flaw in their "way of seeing" (Berger 1972), as one of distance and exteriority rather than one of lived experience. The outsider perspective is useful for tracing patterns and identifying analytical themes, but as Bourdieu (1977) argues, the distant observer cannot understand "practice" as anything other than spectacle. This book was inspired by the idea of exploring landscapes of industrial ruination which have been left behind within an uneven geography of capitalist development (Harvey 1999; Massey 1984; Smith and Harvey 2008). The notion of an uneven geography of capitalist development offers a useful starting point for understanding general patterns and processes of capitalist development, capital flows, and capital abandonment. A strong picture of uneven geographies emerged across all three of the cases, of spatialized socio-economic deprivation and exclusions within landscapes of industrial ruination and adjacent communities. However, even taking Massey's (1984) attention to local contexts into consideration, the model of uneven capitalist development fails to capture the unexpected and the contingent, and tends to bracket social, cultural, and political factors out of the analysis. For example, how can we understand the landscapes of industrial ruination in post-socialist countries, where the relationship with capitalism is entirely different, yet the geography is remarkably similar? How can we account for the role of the state in mediating industrial losses and courting corporate investment? The global geography of industrial ruination is far less tidy than the broad theories of political economy suggest, and instead encompasses complexities and contingencies. Aesthetic and cultural studies of industrial ruins have also proven insufficient for understanding the social and economic processes embedded within industrial ruination. To treat industrial ruins purely as aesthetic objects is to ignore the social relations invested in them, to romanticize them, and to strip them of their meaning and context. Through focusing on the concept of landscape as "dwelling" (Ingold 2000) and conceptualizing industrial ruination as a lived process, this book has bridged dichotomies of culture and economy, form and process, and creation and destruction; and has reconciled perspectives of distance and proximity in considering the complex relationships between landscapes and legacies.

Landscapes

According to theories of political economy and deindustrialization, the logic of capitalist development, whereby capital leaves behind factories and communities in search of cheaper labour and resources, has produced landscapes of industrial decline (Cowie 1999; Harvey 1999; High 2005). Yet none of the cases in this study quite map onto this overarching narrative of deindustrialization. In Niagara Falls, chemical factories left because of a loss of profitability within regional Rust Belt manufacturing markets and because of tightening environmental regulations, rather than a search for cheaper labour and resources. In Newcastle upon Tyne, shipbuilding companies did not abandon the Tyne in search of cheaper labour or resources either: they became less competitive and went bankrupt, were acquired by other companies, or sold their assets to foreign companies and diversified to focus on other industries. Finally, in post-Soviet Ivanovo during the 1990s, the vast number of textile factories producing low-value cotton fabrics could not compete in the new context of the global market economy. Deindustrialization occurred due to profound economic recession, a dramatic decline in domestic demand, and a loss of export competitiveness. However, textile companies worked against the logic of capitalism and gradually re-opened their factories even though they were not profitable, relying on Soviet non-market practices of barter exchanges for short-term survival. These local complexities illustrate the diversity of forms, causes, and responses to deindustrialization, as contrasted with the wider literature that tends to suggest that deindustrialization is a relatively homogenous process linked to the inevitable logic of capital expansion (Bluestone et al. 1981; Harvey 1999; High 2003).

Both cases in Niagara Falls revealed the interrelated themes of unseen qualities and uncertain levels of contamination in the landscape of industrial ruination. Whereas industrial ruination is typically associated with physical landscapes, another type of industrial ruination was present in the contamination of the soil. Invisible and unquantifiable contamination was a particular and pernicious form of industrial ruination in Niagara Falls. The geography of industrial abandonment differed in the Canadian and American contexts, partly because of tighter environmental regulations and corporate liability in Canada, and partly because of the different relationships between industry and tourism on each side. Some old industrial properties have been kept as partially working factories in the United States in order to avoid legal repercussions of hand-over and to delay costs of clean-up. In both cases, governments have been burdened by the costs for clean-up of contaminated sites, which are almost impossible to calculate.

The landscape of ruination in Walker, Newcastle upon Tyne, has two main themes. First, the process of ruination has been painfully prolonged for over thirty years. The end of the shipbuilding era was not properly marked until 2006, decades after its near-total demise. This theme of protracted industrial decline is connected to the second theme: that of regulation. Newcastle City Council and the United Kingdom government have tried to keep shipbuilding and heavy industry afloat, if not economically then at least symbolically, through lifelines and contracts offered to Swan Hunter, and through the support of piecemeal industrial activities along the traditional industrial riverside. The income structure in Newcastle is hourglass-shaped (Bluestone and Harrison 1982), with high-skilled offshore and IT jobs that generate high income, few jobs that produce middle income, and call centre and night economy jobs that generate low income. The key local specificities in Newcastle relate to the interactions between state, community, and capital, with the strong role of the state both in the demise of shipbuilding during the Thatcher years, and its subsequent role in promoting arts- and property-led regeneration.

There were three themes of industrial ruination in Ivanovo: the abundant and pervasive industrial ruins throughout the urban environment; the utilitarian use of space; and the partial reversal of ruination, with a number of partially working, partially abandoned factories. The decline of the textile industry was also due to the fact that it was not competitive in the global market economy. However, instead of a prolonged decline, such as that in Newcastle, the industries in Ivanovo were ruined swiftly and completely. The process is now slowly working in reverse, but depends on barter exchanges rather than market sales (cf. Burawoy et al. 2000; Morrison 2008). Thus, although the landscapes of industrial ruination in both contexts appear to be similar, showing signs of abandonment, near-abandonment, and general decline, they are quite different. These differences can be read in the landscape of Ivanovo through visual evidence of the post-Soviet context (star-topped buildings, Soviet murals, statues and signs, and Soviet architectural style), and visual clues (billowing smoke, security personnel, and vehicles in parking lots) that many factories are starting to function again amidst the abandonment.

A number of factors can account for some of the differences between the local contexts of the three case studies. First, the cases represent different stages in processes of deindustrialization and industrial ruination: recent and continuing (Ivanovo), long and lingering (Niagara Falls), and protracted yet facing regeneration (Walker, Newcastle upon Tyne). A second factor is the role of the state. In Niagara Falls, Ontario, the role of the state was relatively strong, and in Niagara Falls, New York, it was relatively weak, and this discrepancy accounts for differences in environmental law and regulations surrounding corporate accountability. In Walker, the role of the state was the strongest, resulting in high levels of regulation. In Ivanovo, the role of the state had suddenly weakened, which was a factor in the sudden and devastating collapse of industry, although legacies of post-Soviet politics and practices remained embedded within the textile industry.

These three case studies show that landscapes of industrial ruination are not simply barren wastelands, but are deeply interconnected with the social life in wider urban environments. In each of the cases, the residential communities in closest proximity to landscapes of industrial ruination were marked by socio-economic deprivation, urban decline, and social exclusion.

Legacies

Legacies of industrial ruination and urban decline are embodied in local people's experiences, perceptions, and understandings and emerge in unexpected, indirect, or diffuse forms: as uncertainty, as ambivalence, as nostalgia, as trauma, as endurance, and as imagined futures. Legacies do not represent simple nostalgia or straightforward effects of deindustrialization, but rather are the dynamic, temporal dimension of landscapes. Legacies of industrial ruination emerged in each of the cases in relation to lived experiences of its socio-economic and health effects, and to living memory of its sites and processes.

Arguably, the most obvious legacies of deindustrialization are its enduring socio-economic and health implications. However, these legacies were difficult to trace with any precision. A range of quantitative data could support general claims about the negative long-term socio-economic and health impacts of industrial decline, including statistics regarding depopulation, unemployment, epidemiology, deskilling, crime, poverty, and socio-economic deprivation. My research was primarily qualitative, and in trying to connect qualitative with quantitative data, I realized that socio-economic and health issues are quite different matters in terms of hard evidence. Exploring health through qualitative inquiries without the foundation of epidemiological studies to support any claims was difficult in contrast to exploring socio-economic issues qualitatively. Much of the socio-economic data that exists for these case studies is not particularly controversial: job losses and depopulation are known to relate to industrial decline. By contrast, the precise connections between illness and residential proximity to toxic sites remains controversial even within the epidemiological literature. In the case of Niagara Falls, my qualitative investigations pointed towards serious health impacts on residents. However, instead of engaging in what seemed to be a falsely constructed dichotomy between science and what could easily be dismissed as anecdotal or circumstantial evidence, I have not made any claims about the true health impacts of these sites. Instead, I have used my investigations to draw attention to the social impacts of fear, uncertainty, and stress – in combination with health problems that people believe to be related to toxic exposure – that surround the real and perceived health threat of living in proximity to contaminated sites. Indeed, both socio-economic and health impacts of decline and industrial ruination can be framed sociologically, and these connect to lived experiences and memories in the context of uncertainty, disruption, and precariousness.

I also examined the legacies of industrial ruination through social memory. My research showed that social memory is not just about nostalgia or living in the past; it is also about what endures in the present. In this book, I have used the concept of living memory to explore social memory in relation to people's experiences of industrial ruination as a lived process, building on theories of Nora (1989) and Samuel (1994) on the relationship between memory and history. My research offers additional insight into studies of social memory through the particular context of uncertainty in transitional spaces. In all three cases, the places were caught in an uncertain moment between destruction and recreation, reflecting Zukin's notion of an "inner landscape of creative destruction" (1991, 29). Uncertainty and tension between new and old and past and present, relate to both inner and outer landscapes. The industrial past has yet to be left behind in the transition to an undefined post-industrial future, and in the absence of official or unofficial "memorialization" – in the form of monuments, museums, or commemorative gestures – the primary site of social memory is within individuals' stories, as living memory.

People constructed their communities of the past and present in contradictory ways, reflecting differences across generations, social groups, and places. Many of these accounts were nostalgic: people spoke of the bustling commercial activity, tourism, and ample jobs in the heyday of Niagara Falls; people connected to shipbuilding in Walker spoke of the pride and camaraderie associated with making ships; and people spoke about the better days of the vibrant textile industry in Ivanovo. However, many of the accounts expressed continuity with the past, particularly those accounts which emphasized positive features of the present. In Walker, residents proudly regarded their community as something strong enough to endure socio-economic decline. At the same time, they were keenly aware of the social and economic deprivation their existing community faced. In Ivanovo, many residents and workers proudly claimed that the city was still a vibrant textile centre, even as they were aware of its crumbling physical infrastructure. In Highland Avenue, there was continuity through the strength of community spirit, in the face of both contamination and socio-economic decline. In Niagara Falls, Ontario, some remembered the industrial past with ambivalent nostalgia, while those who had suffered the consequences of toxic contamination remembered it as disastrous. In all three cases, people's experiences, perceptions, and memories included a profound sense of uncertainty, disruption, and tension.

Conclusion

These case studies offer not only distinctive conceptual insights, but also have common themes and insights as places caught between the dichotomies of destruction and creation in which the legacies of industrial ruination are fraught, contradictory, and uncertain. There are several key findings and cross-cutting themes within this research:

1. *Industrial ruination is a lived process*. Deindustrialization and industrial ruins are not simply matters of historic record, but represent legacies of industrial ruination: enduring and complex lived realities for people occupying the in-between spaces of post-industrial change.

2. *Reading landscapes of ruination and deprivation*. One can read uneven socio-economic processes of capitalist development within landscapes of industrial ruination and adjacent communities of deprivation through a combination of spatial, visual, and social analysis.
3. *Devastation, but also home*. Many people who live in landscapes of industrial ruination have strong place attachment to their homes and communities despite living among "devastation."
4. *Imagining change, reinventing place*. People's ways of imagining possible futures offer important alternatives to top-down urban planning strategies.
5. *Old industrial centres have diverse challenges and strengths*. The post-industrial model of transformation based on arts- and property-led regeneration and creativity cannot work for all cities within the context of an uneven geography of capitalist development.

Do differences between old industrial cities imply that there is a temporal trajectory of old industrial cities, where Ivanovo represents a recently deindustrialized city, with other cities such as Newcastle further ahead? Does the unevenness of spaces of creation, destruction, consumption, and devastation also increase with the pace of regeneration and development in cities – with relatively widespread deprivation in the city of Ivanovo as a whole, and deprivation clustered in particular areas in cities such as Newcastle? And what of the enduring features of decline in cases where regeneration comes to a halt? How can we address swathes of contaminated land in many cities of the North American Rust Belt? These questions are of concern for all people who are concerned with the future of old industrial cities, in analysing diverse strategies of how to cope with processes of decline and possible renewal. One of the core insights this research reveals is that the one-size-fits-all economic development model of transition from a manufactured-based economy to a post-industrial knowledge- and service-based economy is an insufficient remedy for old manufacturing cities. All three cases show failures in dominant post-industrial visions for solving the problems of industrial decline. However, the diverse stories and perspectives from local people have significant implications for how we might tackle issues of industrial ruination and post-industrial transformation. Places have diverse challenges and strengths, and no single model of economic growth should form the basis for post-industrial renewal.

Notes

Chapter 1

1 Beyond the industrialized West, deindustrialization has occurred in countries throughout the globe, including in areas of Africa, Asia, Latin America, the Middle East, Russia, and Central and Eastern Europe. While a great deal of research has explored global economics of development and transition, particularly in the "newly industrialized countries" of Brazil, Russian, India, and China, there have been limited attempts by Western scholars to account for the different impacts and experiences of deindustrialization around the world. One interesting exception is the case of the Zambian Copperbelt, once described as "the wave of the African future," which flourished in the 1960s only to experience serious deindustrialization in subsequent decades (Ferguson 1999). However, between 2005 and 2007 world copper prices rose, and corporations that had purchased the Zambian copper mines at rock-bottom prices during the late 1990s in the height of industrial decline made considerable profits without reinvesting in community infrastructure (Walsh 2007). This case shows that patterns of industrialization and deindustrialization do not necessarily follow the same timelines around the modern world, and that corporations can be negligent not only during deindustrialization and capital abandonment, but also during re-industrialization.

2 Approaches in contemporary archaeology that attempt to read the materiality of objects and practices from the recent past to learn about socio-economic processes and cultural meanings inspired the methodology in this book. In chapter 5, I engage with these approaches explicitly and draw on a combination of visual, spatial, mobile, and ethnographic methods to read the social and spatial landscapes. The concept of legacies of industrial ruination

also involves a temporal dimension and is another way of excavating the recent past. There are no fixed or clearly defined methods of contemporary archaeology, and my approach borrows insights from this perspective but adopts a reflexive, intuitive, mixed-method case study methodology.

3 I conducted research in Niagara Falls between March and April 2007, which included nineteen in-depth interviews, several group interviews, local archival and documentary material analysis, walking and driving tours with informants, and site and ethnographical observations. In Newcastle upon Tyne, I conducted field research between June 2005 and March 2006, which included thirty qualitative interviews, site and ethnographic observations, walking and driving tours, and document analysis, with five additional follow-up interviews in July 2009. This field work took longer than in the other two cases because it was the pilot case study during which I tested the research design and methodology, and, unlike in the other two case studies, I made a series of trips to the city of three to seven days' duration rather than spending an extended period in the field. I conducted research in Ivanovo between August and September 2006, with follow-up correspondence with key informants in December 2007, and which included eighteen interviews, three group interviews, document analysis, and site and ethnographic observations.

4 According to Yin (1994, 46), the rationale for choosing multiple cases rather than single cases in social research is generally based on "replication" – either literal, in which similar results are predicted, or theoretical, in which contrasting results are predicted but for particular reasons. In the multiple case studies that I selected, I expected to find a combination of Yin's types of replication, literal in the general sense and theoretical in the specific sense.

5 The concept of paradigmatic case studies is derived from Thomas Kuhn's theory of paradigm shifts within scientific thought. Paradigmatic cases "highlight the more general characteristics of the societies in question" (Flyvbjerg 2001, 80).

6 The 2000 English Indices of Deprivation (Noble, Wright, and Dibben 2000) ranked Newcastle at 20 of out 354 Local Authorities in England, where 1 was the most deprived area and 354 the least deprived. The 2004 English Indices of Deprivation (Noble, Wright, and Dibben 2004) analyse multiple deprivation data on the basis of the smaller spatial scale of Super Output Areas (SOAs), rather than cities. The 2004 Indices note that the pattern of severe multiple deprivation in the North East remained similar to the pattern reported in 2000, and was concentrated around old steel, shipbuilding,

and mining areas. The Indices report that 355 of the 10 per cent most deprived SOAs in England are located in the North East.

7 In September 1991, there were riots in the West End of Newcastle related to severe problems of social exclusion in the area, which highlighted the failures of previous regeneration policies. After the riots, the city feared that the area would become a place of lawlessness and anarchy.

Chapter 2

1 In 1980, the Environmental Protection Agency (EPA) released the results of a study of thirty-six residents of the Love Canal area that showed that eleven of the thirty-six people had chromosomal damage of a rare type, which meant an increased risk of miscarriage, stillbirths, cancer, birth defects, and other illnesses (Berton 1992). This report sparked another wave of panic in the Love Canal community and a campaign for permanent relocation. United States President Jimmy Carter eventually signed an agreement for the state to purchase the homes in Love Canal for $15 million, and by February 1981, over 400 families had left the area. For further information, see the following texts: Brown and Clapp 2002, Colten and Skinner 1996, Gibbs 1998, Mazur 1998, and Newman 2003.

2 For example, Vrijheid (2000, 101) flags a problem of quantification within epidemiological studies: "Although a substantial number of studies have been conducted, risks to health from landfill sites are hard to quantify. There is insufficient exposure information and effects of low-level environmental exposure in the general population are by their nature difficult to establish."

3 In 2009, the old downtown was torn up, and a range of theatres, musical events, comedy festivals, restaurants, and boutiques emerged in 2010 and 2011, promoted on a new website (Marconi Consulting Group 2011) under the slogan "Yes this is Downtown Niagara Falls," which contradicts people's memories and expectations of the former derelict downtown.

4 A number of vacant industrial properties are located in this area including the abandoned factories of the American Silver Company, General Abrasives, Canadian Cellucotton Products, several small heavy manufacturing companies, a disused Canadian National Rail yard, and the site of the former chemical factory Cyanamid.

5 Environment Canada's reliance on testing conducted by Cytec consultants rather than independent scientists was one of the core criticisms of the testing and remediation process made by Jim, Unite Here, and other concerned residents.

6 In 1998, as part of the initial brownfield pilot project, the city assessed 5.5 acres of the Union Carbide site. The tests detected some ground contamination and asbestos, and the property was purchased after site "renovations" for $30,000 by Standard Ceramics, a high-tech manufacturer of silicon carbide (Henry 2001). However, this pilot project came to a standstill within a few years because federal and state funding for brownfield redevelopment declined and has only recently (2007) begun to increase.

Chapter 3

1 The commissioning of the sculpture by Gateshead Council in 1994 was criticized heavily in the press because of the great expense, particularly within the regional context of high unemployment. Now it is a major tourist highlight of the North East and is said to be one of the most frequently viewed artworks in the world due to its proximity to the motorway and to passing trains (see Alan 1998).

2 An article in the *Sunday Sun*, dated 24 December 1995 with the headline "Will the last man at Swan Hunter's please turn out the lights," details his story as the last man at the shipyard. In a subsequent article on 14 April 1996, his story reappeared under the headline "Return of the Last Man!"

Chapter 4

1 Our language of communication was in Russian, which limited the depth of our conversations (I had an intermediate level of Russian) but enabled me to conduct research that would have otherwise been difficult on my own.

2 I conducted eighteen interviews and three group interviews in Ivanovo with the help of two student translators and research assistants, Roman and Yuliya, who accompanied me on my interviews.

3 At the time of my research, there was no official tourist office to welcome foreigners to Ivanovo. However, this has since changed. The official Ivanovo city administration website (Ivanovo City Administration 2011) states that the Ivanovo City Tourist Informational Centre had been created as a result of the international project called "Technical support to the development of Ivanovo tourism product" within the framework of TACIS Institution Building Partnership Programme (IBPP) in close cooperation of Ivanovo city administration, the Staffordshire county council and the Centre for Economic and Social Regeneration of Staffordshire University. This suggests that economic development and regeneration through the promotion of tourism has recently been prioritized in Ivanovo through international partnership.

Chapter 5

1 The local politics of community development in Highland and in other communities will be discussed in more detail in chapter 7.

2 Trade union representatives argued that marginal ethnic minority groups or newer migrants to the city are concentrated in the present day area of Glenview-Silvertown and the downtown more generally because of lower housing costs (interview, 5 April 2007), although this trend was not clear from the 2006 Census Figures (Statistics Canada 2006a).

3 The name Voznesensk was dropped in 1932 because it was associated with the religious term "the Ascension" (Vozneseniye).

Chapter 6

1 Newcastle City Council regeneration plans for Walker were originally proposed under a Labour city council in 2000. When the city council changed hands from Labour to the Liberal Democrats in June 2004, the Liberal Democrats picked up on the regeneration plans already in place under Labour. However, after the transition, Labour councillors fought with the Liberal Democrats over the regeneration plans for Walker and sided with residents against demolition, particularly since Walker is traditional Labour territory.

Chapter 7

1 Casinos were banned in large towns and cities in Russia by the national government in January 2007.

2 Walker (1993, 1065) painted a vivid portrait of industrial decline in Łódź just after the fall of the Soviet Union: "The city of Łódź in the 'revolutionary' year of 1989 looked spent, dirty and dishevelled: an urban landscape of decaying textile mills and crumbling late 19th-century facades." Walker remained sceptical about the possibility for Łódź to follow the path of services and new technologies given its low levels of capital accumulation.

References

Alan, Diana, ed. 1998. *Making an Angel*. London: Booth-Clibborn Editions.

Alderson, Arthur S. 1999. "Explaining Deindustrialization: Globalization, Failure or Success?" *American Sociological Review* 64 (5): 701–21.

Altman, Irwin, and S.M. Low, eds. 1992. *Place Attachment, Human Behavior and Environment*. New York: Plenum.

Amin, Ash, ed. 1994. *Post-Fordism: A Reader*. Oxford: Blackwell.

Amin, Ash. 2005. "Local Community on Trial." *Economy and Society* 34 (4): 612–33.

Amin, Ash, and Nigel Thrift. 2002. *Cities: Reimagining the Urban*. Cambridge: Polity.

Anderson, Benedict R. 1991. *Imagined Communities: Reflections on the Origin and Spread of Nationalism*. Revised and extended edition. London: Verso.

Anderson, Jon. 2004. "Talking whilst Walking: A Geographical Archaeology of Knowledge." *Area* 36 (3): 254–61.

Ashwin, Sarah, and Simon Clarke, eds. 2003. *Russian Trade Unions and Industrial Relations in Transition*. Basingstoke: Palgrave Macmillan.

Audit Commission. 2009. *Neighbourhood Renewal*. London: Audit Commission.

Bailey, Christopher, Steven Miles, and Peter Stark. 2004. "Culture-led Urban Regeneration and the Revitalisation of Identities in Newcastle, Gateshead and the North East of England." *International Journal of Cultural Policy* 10 (1): 47–65.

Banks, Marcus. 2001. *Visual Methods in Social Research*. London: Sage.

Bell, Colin, and Howard Newby. 1971. *Community Studies: An Introduction to the Sociology of the Local Community*. London: Allen & Unwin.

Beck, Ulrich. 1992. *Risk Society: Towards a New Modernity*. London and Thousand Oaks: Sage.

Bell, Daniel. 1973. *The Coming of Post-industrial Society: A Venture in Social Forecasting*. New York: Basic Books.

Benjamin, Walter. 2000. *The Arcades Project*. Cambridge: The Belknap Press of Harvard University Press.

Benjamin, Walter. 1999. *Illuminations*. Edited by Hannah Arendt. Translated by H. Zorn. New York: Pimlico. Original edition 1968.

Berger, John. 1972. *Ways of Seeing*. London: British Broadcasting Corporation and Penguin Books.

Berketa, Rick. 2005. *Niagara Falls Environmental Impact*. Accessed 13 March 2005. http://www.iaw.com/~falls/environment.html.

Berman, Marshall. 1983. *All that is Solid Melts into Air: The Experience of Modernity*. London: Verso.

Berton, Pierre. 1992. *Niagara: A History of the Falls*. Toronto: McClelland & Stewart.

Blokland, T. 2001. "Bricks, Mortar, Memories: Neighbourhood and Networks in Collective Acts of Remembering." *International Journal of Urban and Regional Research* 25 (2): 268–83.

Bluestone, Barry, and Bennett Harrison. 1982. *The Deindustrialization of America: Plant Closings, Community Abandonment, and the Dismantling of Basic Industries*. New York: Basic Books.

Bourdieu, Pierre. 1977. "Outline of a Theory of Practice." *Cambridge Studies in Social Anthropology* 16. Cambridge: Cambridge University Press.

Bourdieu, Pierre. 1999. "Site Effects." In *The Weight of the World*, edited by Pierre Bourdieu et al., translated by Priscilla P. Ferguson, 123–9. Stanford: Stanford University Press.

Boyer, M. Christine. 1994. *The City of Collective Memory: Its Historical Imagery and Architectural Entertainments*. Cambridge: MIT Press.

Boyer, Robert, and Jean-Pierre Durand. 1997. *After Fordism*. Basingstoke: Macmillan Business.

Brent, Jeremy. 2009. *Searching For Community: Representation, Power and Action on an Urban Estate*. Bristol: Policy Press.

Briggs, Asa. 1968. *Victorian Cities*. Harmondsworth: Penguin Books.

Brown, Barbara, Douglas Perkins, and Graham Brown. 2003. "Place Attachment in a Revitalizing Neighborhood: Individual and Block Levels of Analysis." *Journal of Environmental Psychology* 23 (3): 259–71.

Brown, Phil. 2002. *Health and the Environment*. Annals of the American Academy of Political and Social Science, v. 584. Thousand Oaks: Sage.

Brownfield Opportunity Area Program, The. 2008. "The Highland Avenue Neighborhood, Niagara Falls New York." http://www.shapehighlandsfuture.com.

Browning, Genia. 1992. "The Zhensovety Revisited." In *Perestroika and Soviet Women*, edited by Mary Buckley, 97–117. Cambridge: Cambridge University Press.

Buchli, Victor, ed. 2002. *The Material Culture Reader.* Oxford: Berg.

Buchli, Victor, and Gavin Lucas, eds. 2001. *Archaeologies of the Contemporary Past*. London: Routledge.

Bullard, Robert D. 1993. *Confronting Environmental Racism: Voices from the Grassroots*. Boston: South End Press.

Burawoy, Michael, and Katherine Verdery, eds. 1999. *Uncertain Transition: Ethnographies of Change in the Postsocialist World*. Lanham: Rowman & Littlefield.

Burawoy, Michael, Pavel Krotov, and Tatyana Lytkina. 2000. "Involution and Destitution in Capitalist Russia." *Ethnography* 1 (1): 43–65.

Büscher, M., and J. Urry. 2009. "Mobile Methods and the Empirical." *European Journal of Social Theory* 12 (1): 99–116.

Byrne, David. 1999. *Social Exclusion*. Birmingham and Philadelphia: Open University Press.

Byrne, David. 2000. "Newcastle's Going for Growth: Governance and Planning in a Postindustrial Metropolis." *Northern Economic Review* 30: 3–16.

Cameron, Stuart. 2003. "Gentrification, Housing Redifferentiation and Urban Regeneration: 'Going for Growth' in Newcastle upon Tyne." *Urban Studies* 40 (12): 2367–82.

Cameron, Stuart, and J. Coaffee. 2004. "Art, Gentrification and Regeneration: From Artist as Pioneer to Public Arts." ENHR Conference Paper, Cambridge.

Carpenter, Mick, Belinda Freda, and Stuart Speeden. 2007. *Beyond the Workfare State: Labour Markets, Equalities and Human Rights*. Bristol: Policy.

Charles, David, and P. Benneworth. 2001. "Situating the North East in the European space economy." In *A Region in Transition: North East England at the Millennium*, edited by John Tomaney and Neil Ward. Burlington: Ashgate.

Chossudovsky, Michel. 2003. *The Globalization of Poverty and the New World Order.* 2nd ed. Shanty Bay: Global Outlook.

City of Niagara Falls. 2011. "Niagara Falls USA." http://www.niagarafallsusa.org.

Clairmont, Donald H., and Dennis William Magill. 1974. *Africville: The Life and Death of a Canadian Black Community*. Toronto: McClelland and Stewart.

Clifford, James. 1997. *Routes: Travel and Translation in the Late Twentieth Century*. Cambridge: Harvard University Press.

Cochrane, A. 2006. *Understanding Urban Policy: A Critical Approach.* Oxford: Blackwell.

Cole, Luke W., and Sheila R. Foster. 2001. *From the Ground Up: Environmental Racism and the Rise of the Environmental Justice Movement*. New York: New York University Press.

Colten, Craig E., and Peter N. Skinner. 1996. *The Road to Love Canal: Managing Industrial Waste Before EPA*. Austin: University of Texas Press.

Connerton, Paul. 1989. *How Societies Remember, Themes in the Social Sciences*. Cambridge: Cambridge University Press.

Cooper, Keith. 2011. "Pathfinder areas plea for £60m lifeline." *Inside Housing*, 12 June. http://www.insidehousing.co.uk.

Couch, Chris, C. Fraser, and S. Percy, eds. 2003. *Urban Regeneration in Europe, Real Estate Issues*. Oxford: Blackwell.

Cowie, Jefferson. 1999. *Capital Moves: RCA's 70-Year Quest for Cheap Labour*. Ithaca: Cornell University Press.

Cowie, Jefferson, and J. Heathcott, eds. 2003. *Beyond the Ruins: The Meanings of Deindustrialization*. Ithaca: ILR Press.

Coyle, Diane. 1998. "Britain's Urban Boom: The New Economics of Cities." Working Paper, No. 7. London: Comedia in association with Demos.

Cresswell, Tim. 1996. *In Place/Out of Place: Geography, Ideology, and Transgression*. Minneapolis: University of Minnesota Press.

Cresswell, Tim. 2004. *Place: A Short Introduction*. Oxford: Blackwell.

Crow, Graham. 2002. "Community Studies: Fifty years of Theorization." *Sociological Research Online* 7 (3): http://www.socresonline.org.uk/7/3/crow.html.

Cytec Industries. 2009. "Company Overview." Last modified 2009. Accessed 1 March 2011. http://www.cytec.com.

Darby, H.C. 1962. "The Problem of Geographical Description." *Transactions of the Institute of British Geographers* 30: 1–14.

Davis, Fred. 1979. *Yearning for Yesterday: A Sociology of Nostalgia*. New York: Free Press.

Davis, Mike. 1990. *City of Quartz: Excavating the Future in Los Angeles*. London: Verso.

de Certeau, Michel 1984. *The Practice of Everyday Life*. Berkeley: University of California Press.

Dehejia, Vivek H. 1997. *Will Gradualism Work When Shock Therapy Doesn't?* London: Centre for Economic Policy Research.

Dench, Geoff, Kate Gavron, and Michael Young. 2006. *The New East End: Kinship, Race and Conflict*. London: Profile Books.

Dicken, Peter. 2002. "Global Manchester: From Globaliser to Globalised." In *City of Revolution: Restructuring Manchester*, edited by J. Peck and K. Ward, 18–33. Manchester: Manchester University Press.

Doel, Marcus, and Phil Hubbard. 2002. "Taking World Cities Literally: Marketing the City in a Global Space of Flows." *City* 6 (3): 351–68.

Doll, Richard, and A. McLean. 1979. *Long-term Hazards from Environmental Chemicals: A Royal Society Discussion*. London: Royal Society.

Donald, James. 1999. *Imagining the Modern City*. London: Athlone.

Dougan, David. 1968. *The History of North East Shipbuilding*. London: Allen and Unwin.

Draper, William M. 1994. "Environmental Epidemiology: Effects of Environmental Chemicals on Human Health." *Advances in Chemistry Series* 241.

Dubinsky, Karen. 1999. *The Second Greatest Disappointment: Honeymooning and Tourism at Niagara Falls*. New Brunswick: Rutgers University Press.

Dudley, Kathryn Marie. 1994. *The End of the Line: Lost Jobs, New Lives in Postindustrial America*. Chicago: University of Chicago Press.

Dunn, Clara, and W. Dunn. 1998. "Brownfields in Niagara Falls: City Prepares for Future Redevelopment of the Highland Avenue Neighborhood." *Niagara Falls: Moving Towards Tomorrow*, Fall: 2.

Edensor, Tim. 2005. *Industrial Ruins: Space, Aesthetics and Materiality*. Oxford: Berg.

Emmison, Michael, and Philip Smith. 2000. *Researching the Visual*. London: Sage.

Fentress, James, and C. Wickham. 1992. *Social Memory, New Perspectives on the Past*. Oxford: Blackwell.

Fletcher, Richard. 2010. "Last two Swan Hunter cranes come down," *Evening Chronicle*, Newcastle upon Tyne, June 5. http://www.chroniclelive.co.uk/north-east-news/evening-chronicle-news.

Florida, Richard L. 2005. *Cities and the Creative Class*. New York: Routledge.

Flyvbjerg, Bent. 2001. *Making Social Science Matter: Why Social Inquiry Fails and How it Can Count Again*. New York: Cambridge University Press.

Freeland, Chrystia. 2000. *Sale of the Century: The Inside Story of the Second Russian Revolution*. London: Little Brown.

French, Ron, and Ken Smith. 2004. *Lost Shipyards of the Tyne*. Newcastle upon Tyne: Tyne Bridge Publishing.

Fried, M. 1963. "Grieving for a Lost Home." In *The Urban Condition: People and Policy in the Metropolis*, edited by L.J. Duhl. New York: Basic Books.

Fried, M. 2000. "Continuities and Discontinuities of Place." *Journal of Environmental Psychology* 20 (3): 193–205.

Gaffikin, Frank, and M. Morrissey. 1999. *City Visions: Imagining Place, Enfranchising People*. London: Pluto Press.

Gerber, T.P., and O. Mayorova. 2006 "Dynamic Gender Relations in a Post-Socialist Labor Market: Russia, 1991-1997." *Social Forces* 84 (4): 2047–75.

Gibbs, Lois. 1998. *Love Canal: The Story Continues*. Twentieth anniversary revised edition. Gabriola Island: New Society Publishers.

Giddens, Anthony. 1991. *Modernity and Self-Identity: Self and Society in the Late Modern Age*. Cambridge: Polity.

Gilroy, Paul. 1993. *The Black Atlantic: Modernity and Double Consciousness*. London: Verso.

González, Sara. 2011. "Bilbao and Barcelona 'in Motion': How Urban Regeneration 'Models' Travel and Mutate in the Global Flows of Policy Tourism." *Urban Studies* 48 (7): 1397–418.

Grindea, Dan. 1997. *Shock Therapy and Privatization: An Analysis of Romania's Economic Reform*. East European Monographs, no. 479. New York: East European Monographs; Distributed by Columbia University Press.

Gustafson, P. 2001. "Roots and Routes: Exploring the Relationship between Place Attachment and Mobility." *Environment and Behavior* 33 (5): 667–86.

Halbwachs, Maurice. 1980. *The Collective Memory*. New York: Harper & Row.

Hall, Tim, and Phil Hubbard, eds. 1998. *The Entrepreneurial City: Geographies of Politics, Regime and Representation*. Chichester: Wiley.

Hang, Walter, and Joseph Salvo. 1981. *The Ravaged River: Toxic Chemicals in the Niagara. Niagara Falls*. New York: The New York Public Interest Research Group.

Harper, Ralph. 1966. *Nostalgia: An Existential Exploration of Longing and Fulfilment in the Modern Age*. Cleveland: Press of Western Reserve University.

Harvey, David. 1989. *The Condition of Postmodernity: An Enquiry into the Origins of Cultural Change*. Oxford: Basil Blackwell.

Harvey, David. 1999. *The Limits to Capital*. Verso edition. New York: Verso.

Harvey, David. 2000. *Spaces of Hope*. Berkeley: University of California Press.

Harvey, David. 2003. "Contested Cities: Social Process and Spatial Form." In *The City Reader*, edited by R.T. LeGates and F. Stout. New York: Routledge.

Healy, Robert. 2006. "The Common's Problem and Canada's Niagara Falls." *Annals of Tourism Research* 33 (2): 525–44.

Henry, Sherryl. 2001. "For Developers in Niagara Falls, the Honeymoon's Just Beginning." In *Brownfield Success Stories*. Niagara Falls: United States Environmental Protection Agency June, EPA 500-F-01-224, http://epa.gov/brownfields/success/success_archives.htm.

Hess, Daniel Baldwin. 2005. "Access to Employment for Adults in Poverty in the Buffalo–Niagara Region." *Urban Studies* 42 (7): 1177–200.

High, Steven. 2003. *Industrial Sunset: The Making of North America's Rust Belt, 1969-1984*. Torónto: University of Toronto Press.

High, Steven. 2007. *Corporate Wasteland: The Landscape and Memory of Deindustrialization*. Ithaca: Cornell University Press.

Hilditch, Peter J. 1990. "Defence Procurement and Employment: The Case of UK Shipbuilding." *Cambridge Journal of Economics* 14 (1): 483–96.

Hoffman, Andrew J. 1999. "Institutional Evolution and Change: Environmentalism and the U.S. Chemical Industry." *The Academy of Management Journal* 42 (4): 351–71.

Hollands, Robert, and P. Chatterton. 2002. "Changing Times for an Old Industrial City: Hard Times, Hedonism and Corporate Power in Newcastle's Nightlight." *City* 6 (3): 291–315.

Hudson, Ray. 1998. "Restructuring Region and State: The Case of North East England." *Tijdschrift voor Economische en Sociale Geografie* 89 (1): 15–30.

Hurdley, Rachel. 2006. "Dismantling Mantelpieces: Narrating Identities and Materializing Culture in the Home." *Sociology* 40 (4): 717–33.

Ingold, Tim. 2000. *The Perception of the Environment: Essays on Livelihood, Dwelling and Skill*. London: Routledge.

Ingram, Paul, and C. Inman. 1996. "Institutions, Intergroup Competition, and the Evolution of Hotel Populations around Niagara Falls." *Administrative Science Quarterly* 41 (4): 629–58.

Irwin, William. 1996. *The New Niagara: Tourism, Technology, and the Landscape of Niagara Falls, 1776–1917*. University Park: Pennsylvania State University Press.

Ivanovo City Administration. 2011. "Ivanovo City Administration." Accessed March 2011. http://www.ivanovo.ru.

Jakle, John A., and David Wilson. 1992. *Derelict Landscapes: The Wasting of America's Built Environment*. Savage: Rowman and Littlefield Publishers.

Januarius, J. 2009. "Feeling at Home: Interiors, Domesticity, and the Everyday Life of Belgian Limburg Miners in the 1950s." *Home Cultures* 6 (1): 43–70.

Jessop, Bob. 1991. *Fordism and Post-Fordism: A Critical Reformulation*. Lancaster: Lancaster Regionalism Group.

Josephson, Julian. 1983. "Exposure to Chemical Waste Sites." *Environmental Science and Technology* 17 (7): 286–9.

Kaczmarek, Sylwia, and C. Young. 1999. "Changing the Perception of the Post-socialist City: Place Promotion and Imagery in Łódź, Poland." *The Geographical Journal* 165: 183–91.

Kennedy, Rob. 2006. "Yard Bosses Criticised for Site Neglect." *Evening Chronicle*, Newcastle upon Tyne, June 1. http://www.chroniclelive.co.uk/north-east-news.

Knowles, Caroline, and Paul Sweetman, eds. 2004. *Picturing the Social Landscape: Visual Methods and the Sociological Imagination*. London and New York: Routledge.

Kouznetsov, Alexei. 2004. "Russian Old-Industry Regions in the Transformation Process." In *Ivanovo: Shrinking Cities Working Paper*, edited by S. Sitar and A. Sverdlov, 31–40. Ivanovo: Federal Cultural Foundation.

Landry, Charles, and F. Bianchini. 1995. *The Creative City*. London: Demos Comedia.

Lane, Paul. 2010. "Locals Still Feel Love Canal Effect." *Tonawanda News*, April 27.

Lassiter, Luke Eric, et al. 2005. *The Other Side of Middletown: Exploring Muncie's African American Community*. Walnut Creek, California: AltaMira Press.

Laz, Anneliese, and P. Laz. 2001. "Imaginative Landscapes out of Industrial Dereliction." In *Cities for the New Millennium*, edited by M. Echenique and A. Saint, 73–78. London: Spon Press.

Leadbetter, C. 1998 "Who Will Own the Knowledge Economy?" *The Political Quarterly* 69 (4): 375–85.

Ledwith, Margaret. 2005. *Community Development: A Critical Approach*. Bristol: Policy Press.

Lefebvre, Henri. 1991. *The Production of Space*. Oxford: Blackwell.

Lefebvre, Henri, and Robert Bononno. 2003. *The Urban Revolution*. Minneapolis: University of Minnesota Press.

Lipietz, Alain. 1992. *Towards a New Economic Order: Postfordism, Ecology and Democracy*. New York: Polity.

LiPuma, Edward, and T. Koelble. 2005. "Cultures of Circulation and the Urban Imaginary: Miami as Example and Exemplar." *Public Culture* 17 (1): 153–79.

Liszewski, Stanislaw, C. Young, and B. Gontar. 1997. *A Comparative Study of Łódź and Manchester: Geographies of European Cities in Transition*. Łódź: University of Łódź Press.

Long, Robert, and Wendy Wilson. 2011. *Housing Market Renewal Pathfinders Parliamentary Note*. London: House of Commons.

Lorenz, Edward H. 1991. "An Evolutionary Explanation for Competitive Decline: The British Shipbuilding Industry, 1890–1970." *The Journal of Economic History* 51 (4): 911–35.

Loroque, Corey. 2007. "Councils Votes to Build Four-Pad Complex." *Niagara Falls Review*, May 7. http://www.niagarafallsreview.ca/Default.aspx.

Low, Setha M. 1992. "Symbolic Ties that Bind: Place Attachment in Plaza." In *Place Attachment*, edited by I. Altman and S.M. Low, 165–85. New York: Plenum.

MacDonald, Mott. 2005. *Walker Riverside Area Action Plan: Draft Sustainability Appraisal Scoping Report*. Newcastle upon Tyne: Newcastle City Council.

MacKenzie, Maxwell 2001. *American Ruins: Ghosts on the Landscape*. Afton: Afton Historical Society Press.

Madanipour, Ali. 1998. "Social Exclusion and Space." In *Social Exclusion in European Cities: Processes, Experiences and Responses*, edited by A. Mandanipour, G. Cars, and J. Allen, 75–94. London: Jessica Kingsley.

Madanipour, Ali, and M. Bevan. 1999. *Walker, Newcastle upon Tyne: A Neighbourhood in Transition*. CREUE Occasional Papers Series (No. 2). Newcastle upon Tyne: University of Newcastle upon Tyne.

Mallett, S. 2004. "Understanding Home: A Critical Review of the Literature." *Sociological Review* 52 (1): 62–89.

Malysheva, Yulia. 2007. "A Revolution of the Mind." *Demokratizatsiya: The Journal of Post-Soviet Democratization* 15 (1): 117–28.

Marconi Consulting Group. 2011. "Yes this is Niagara Falls." Accessed February 2011. URL no longer valid. Replacement URL accessed February 2012. http://www.marconiconsulting.com.

Marcuse, Peter. 1997. "The Enclave, the Citadel, and the Ghetto: What has Changed in the Post-Fordist U.S. City." *Urban Affairs Review* 33 (2): 228–64.

Massey, Doreen. 1994. *Space, Place and Gender*. Minneapolis: University of Minnesota Press.

Massey, Doreen. 1984. *Spatial Divisions of Labour: Social Structures and the Geography of Production*. Basingstoke: MacMillan.

Mazur, Allan. 1998. *A Hazardous Inquiry: The Rashomon Effect at Love Canal*. Cambridge: Harvard University Press.

McDowell, Linda. 1999. *Gender, Identity and Place: Understanding Feminist Geographies*. Cambridge: Polity.

McGreevy, Patrick, and C. Merritt. 1991. *The Wall of Mirrors: Nationalism and Perceptions of the Border at Niagara Falls*, Borderlands Monograph Series; 5. Orono: Borderlands Project.

McMillan, Paul. 2006. "Pet Ship Boys." *Evening Chronicle*, Newcastle upon Tyne, March 3. http://www.chroniclelive.co.uk/north-east-news.

Merleau-Ponty, Maurice, and C. Smith. 1989. *Phenomenology of Perception*. London: Routledge.

Milkman, Ruth. 1997. *Farewell to the Factory: Auto Workers in the Late Twentieth Century*. Berkeley: University of California Press.

Miller, Daniel. 2006. "Things that Bright up the Place." *Home Cultures* 3 (3): 235–50.

Mills, C. Wright. 1959. *The Sociological Imagination*. New York: Oxford University Press.

Mitman, Gregg, M. Murphy, and C. Sellers. 2004. *Landscapes of Exposure: Knowledge and Illness in Modern Environments*. Chicago: University of Chicago Press.

Moretti, Luke. 2007. "Life Inside Laborers' Local 91." *News 4*, February 2. http://www.wivb.com/Global/story.asp?S=6029698.

Morrison, Claudio. 2008. *A Russian Factory Enters the Market Economy*. London and New York: Routledge.

Morrison, Doug. 2007. "Urban Regeneration – Finding a Place to Grow." *Global City Magazine*, 2 March.

Mumford, Katherine, and Ann Power. 2003. *East Enders: Family and Community in East London*. Bristol: Policy Press.

Murray, Gordon. 1992. "Cyanamid and the City Share History." *Niagara Falls Review*, February 1.

Nayak, Anoop. 2003. *Race, Place and Globalization: Youth Cultures in a Changing World*. Oxford: Berg.

Newcastle City Council. 2007. "Walker Riverside." Accessed December 2007. URL no longer valid. http://www.walker-riverside.co.uk/living.html.

Newcastle City Council. 2008. *Newcastle in 2021: A Regeneration Strategy for Newcastle*. Newcastle upon Tyne: Newcastle City Council.

Newcastle City Council. 2009. "Walker Riverside." Accessed July 2009. URL no longer valid. http://www.walker-riverside.co.uk/homes.html.

Newman, Richard. 2003. "From Love's Canal to Love Canal." In *Beyond the Ruins: The Meanings of Deindustrialization*, edited by J. Cowie and J. Heathcott, 112–38. Ithaca: Cornell University Press.

Niagara Falls Empire Zone. 2007. "Brownfields." Last modified 2007. Accessed April 2007. http://www.nfez.org/planning/programs/brownfields.htm.

Niagara Falls Review. 2005. "No Link between Cyanamid and Residents' Health – Findings of a Report by Region." *Niagara Falls Review*, February 17.

Niagara Region. 2010. "Niagara Region Neighbourhood Profiles." Thorold: Niagara Region. Accessed 4 February 2011. http://www.niagararegion.ca/social-services/addressing-poverty-in-niagara.aspx.

Noble, Michael, Gemma Wright, and Chris Dibben. 2000. *The English Indices of Deprivation 2000*. London: Office of the Deputy Prime Minister.

Noble, Michael, Gemma Wright, and Chris Dibben. 2004. *The English Indices of Deprivation 2004* (revised). London: Office of the Deputy Prime Minister.

Noble, Michael, et al. 2007. "The English Indices of Deprivation 2007: Summary." In *The English Indices of Deprivation 2007*. London: Communities and Local Government.

Nora, Pierre. 1989. "Between Memory and History." *Representations* 26 (1): 7–24.

O'Connor, Justin. 1998. "Popular Culture, Cultural Intermediaries and Urban Regeneration." In *The Entrepreneurial City: Geographies of Politics, Regime and Representation*, edited by T. Hall and P. Hubbard, 225–41. Chichester: Wiley.

Oushakine, Serguei. 2000. "In the State of Post-Soviet Aphasia: Symbolic Development in Contemporary Russia." *Europe-Asia Studies* 52 (6): 991–1016.

Oushakine, Serguei. 2007. "'We're Nostalgic but We're not Crazy': Retrofitting the Past in Russia." *The Russian Review* 66 (3): 451–82.

Pauli, Lori, et al. 2003. *Manufactured Landscapes: The Photographs of Edward Burtynsky*. Ottawa: National Gallery of Canada in association with Yale University Press.

Peck, Jamie. 2001. *Workfare States*. New York: Guilford Press.

Peck, Jamie, and Kevin Ward. 2002. *City of Revolution: Restructuring Manchester*. Manchester: Manchester University Press.

Pelligrini, Jennifer. 2005. "Resident Questions Cyanamid Safety." *Niagara Falls Review*, October 28.

Persky, Joseph, and Wim Wiewel. 2000. *When Corporations Leave Town: The Costs and Benefits of Metropolitan Job Sprawl*. Detroit: Wayne State University Press.

Pink, Sarah. 2001. *Doing Visual Ethnography*. London: Sage.

Quilley, Stephen. 1999. "Entrepreneurial Manchester: The Genesis of Elite Consensus." *Antipode* 31 (2): 185–211.

Ragin, Charles C. 1987. *The Comparative Method*. Berkeley: University of California Press.

Rae, Ian, and K. Smith. 2001. *Swan Hunter: The Pride and the Tears*. Newcastle upon Tyne: Tyne Bridge Publishing.

RCI Consulting. 2006. *Pilot Project Area Study: Brownfield Community Improvement Plan*. Niagara Falls: City of Niagara Falls, Ontario.

Relph, Edward. 1976. *Place and Placelessness*. London: Pion.

Remington, Thomas F. 2006. *Politics in Russia*. Fourth edition. New York: Pearson-Longman.

Reuters. 2007. "Bharati Shipyard Buys UK's Swan Hunter." *Reuters*, April 9. http://www.reuters.com/article/2007/04/10/idUSBOM19806220070410.

Ricciuto, Tony. 1995. "Cyanamid – Memories of Swimming and Fries." *Niagara Falls Review*, March 31.

Richardson, Ranald, V. Belt, and N. Marshall. 2000. "Taking Calls to Newcastle: The Regional Implications of the Growth in Call Centres." *Regional Studies* 34 (4): 357–69.

Roberts, Ian. 1992. *Craft, Class and Control: The Sociology of a Shipbuilding Community*. Edinburgh: Edinburgh University Press.

Roberts, Ian. 2007. "Collective Representations, Divided Memory and Patterns of Paradox: Mining and Shipbuilding." *Sociological Research Online* 12 (6). http://www.socresonline.org.uk/12/6/6.html.

Robinson, Fred. 2002. "The North East: A Journey through Time." *City* 6 (3): 317–34.

Robinson, Fred. 2005. "Regenerating the West End of Newcastle: What Went Wrong?" *Northern Economic Review* 36 (Summer): 15–41.

Room, Robin, N. Turner, and A. Ialomiteanu. 1999. "Community Effects of the Opening of the Niagara Casino." *Addiction* 94 (10): 1449–66.

Rose, Gillian. 1993. *Feminism and Geography: The Limits of Geographical Knowledge*. Cambridge: Polity Press.

Rose, Gillian. 2007. *Visual Methodologies*. Second edition. Thousand Oaks: Sage.

Rowthorn, Bob, and R. Ramaswamy. 1997. "Deindustrialization: Causes and Implications." IMF Working Paper. Washington: International Monetary Fund Research Dept.

Russian Federal Service of State Statistics. 2002. *The All-Russia Census of Population*. Moscow: Government of the Russian Federation.

Sadler, Colin. 2006. "Death Toll from Asbestos is Still Going Up." *Evening Chronicle*, Newcastle upon Tyne, March 15. http: //www.chroniclelive.co.uk/north-east-news.

Samuel, Raphael. 1994. *Theatres of Memory*. London: Verso.

Sargin, G. 2004. "Displaced Memories, or the Architecture of Forgetting and Remembrance." *Environment and Planning D: Society and Space* 22 (5): 659–80.

Sassen, Saskia. 2002. *Global Networks, Linked Cities*. New York; London: Routledge.

Savage, K. 2003. "Monuments of a Lost Cause: The Postindustrial Campaign to Commemorate Steel." In *Beyond the Ruins: the Meanings of Deindustrialization*, edited by J. Cowie and J.C. Heathcott, 237–58. Ithaca: ILR Press.

Schneekloth, Lynda, and Robert Shibley. 2005. "Imagine Niagara." *Journal of Canadian Studies* 39 (3): 105–20.

Schumpeter, Joseph A. 1965. *Capitalism, Socialism and Democracy*. Fourth edition, University books. London: Allen & Unwin.

Schwirtz, Michael. 2007. "Russia's Political Youths." *Demokratizatsiya: The Journal of Post-Soviet Democratization* 15 (1): 73–85.

Sen, Amartya. 2009. *The Idea of Justice*. London and New York: Allen Lane.

Sennett, Richard. 1998. *The Corrosion of Character: The Personal Consequences of Work in the New Capitalism*. New York: Norton.

Shackel, Paul A., and M. Palus. 2006. "Remembering an Industrial Landscape." *International Journal of Historical Archaeology* 10 (1): 49–71.

Shaw, Christopher, and M. Chase, eds. 1989. *The Imagined Past: History and Nostalgia*. Manchester: Manchester University Press.

Shields, Rob. 1991. *Places on the Margin: Alternative Geographies of Modernity*. London: Routledge.

Shushpanov, A, and S. Oleksenko. 2005. *Ivanovo: City of Brides*. Ivanovo: Ivanovo City Administration.

Simmel, Georg. 1997. *Simmel on Culture: Selected Writings, Theory, Culture & Society* (unnumbered), edited by David Frisby and Mike Featherstone. London: Sage.

Sinclair, Iain. 2002. *London Orbital*. London: Penguin.

Sitar, Sergei, and A. Sverdlov, eds. 2004. *Ivanovo: Shrinking Cities Working Paper*. Ivanovo: Federal Cultural Foundation.

Skeffington, Mark. 1992. "Cyanamid Niagara Closes: 243 Jobless by Year's End." *Niagara Falls Review*, March 5.

Skeffington, Mark. 1993. "The Cyanamid Closing – A Year Later." *Niagara Falls Review*, April 10.

Smith, Neil. 1984. *Uneven Development: Nature, Capital and the Production of Space*. Oxford: Blackwell.

Smith, Neil. 2002. "New Globalism, New Urbanism: Gentrification as Global Urban Strategy." *Antipode* 34 (3): 427–50.

Smith, Neil, and D. Harvey. 2008. *Uneven Development: Nature, Capital, and the Production of Space*. Athens: University of Georgia Press.

Soja, Edward. 1996. *Thirdspace: Journeys to Los Angeles and Other Real-and-imagined Places*. Cambridge: Blackwell.

Soja, Edward. 2000. *Postmetropolis: Critical Studies of Cities and Regions*. Oxford: Blackwell Publishers.

Stacey, Margaret. 1969. "The Myth of Community Studies." *The British Journal of Sociology* 20 (2): 134–47.

Statistics Canada. 2006a. *Neighbourhood Profiles: Niagara Region*. In Semi-custom Profile of Selected Geographic Areas in Niagara Region, 2006 Census. Ottawa: Statistics Canada.

Statistics Canada. 2006b. *Census Tract (CT) Profiles, 2006 Census*. Accessed 5 February 2011. http://www12.statcan.ca/census-recensement/2006/dp-pd/prof/92-597.

Staudohar, Paul D., and H.E. Brown. 1987. *Deindustrialization and Plant Closure*. Lexington: D.C. Heath.

Stephan, Johannes. 1999. *Economic Transition in Hungary and East Germany: Gradualism and Shock Therapy in Catch-Up Development*. New York: St Martin's Press.

Stewart, Kathleen. 1996. *A Space on the Side of the Road: Cultural Poetics in an "Other" America*. Princeton: Princeton University Press.

Summers, Bob. 1971. "Blacks at Carbide are Charging Bias." *Niagara Falls Gazette*, October 31.

Tallon, A. 2010. *Urban Regeneration in the UK*. London and New York: Routledge.

Tesh, Sylvia N. 1993. "Environmentalism, Pre-environmentalism, and Public Policy." *Policy Sciences* 26 (1): 1–20.

Tomaney, John, and N. Ward, eds. 2001. *A Region in Transition: North East England at the Millennium*, Urban and Regional Planning and Development. Aldershot: Ashgate.

Treivish, Andrei. 2004. "Ivanovo Long-Term Socio-economic and Urban Development." In *Ivanovo: Shrinking Cities Working Paper*, edited by S. Sitar and A. Sverdlov, 11–27. Ivanovo: Federal Cultural Foundation.

Trigg, Dylan. 2009. "The Place of Trauma: Memory, Hauntings and the Temporality of Ruins." *Memory Studies* 2 (1): 87–101.

Tuan, Yi-fu. 1977. *Space and Place: The Perspective of Experience*. London: Edward Arnold.

Turner, Nigel, A. Ialomiteanu, and R. Room. 1999. "Checkered Expectations: Predictors of Approval of Opening a Casino in the Niagara Community." *Journal of Gambling Studies* 15 (1): 45–70.

Urry, J. 2007. *Mobilities*. Cambridge: Polity Press.

US Census Bureau. 2011. "State and County QuickFacts: Niagara Falls, New York," updated 8 July 2009. Accessed 4 February 2011. http://quickfacts.census.gov/qfd/states/36/3651055.html.

Van der Hoorn, Mélanie. 2003. "Exorcizing Remains: Architectural Fragments as Intermediaries between History and Individual Experience." *Journal of Material Culture* 8 (2): 189–213.

Vergara, Camilo José. 1999. *American Ruins*. New York: The Monacelli Press.

Vrijheid, Martine. 2000. "Health Effects of Residence near Hazardous Waste Landfill Sites: A Review of Epidemiologic Literature." *Environmental Health Perspectives* 108 (1): 101–12.

Wacquant, Loïc J.D. 1999. "America as Social Dystopia: The Politics of Urban Disintegration or the French Uses of the 'American Model'." In *The Weight of the World*, edited by Pierre Bourdieu et al., translated by Priscilla P. Ferguson, 130–9. Stanford: Stanford University Press.

Walker, Antony. 1993. "Łódź: The Problems Associated with Restructuring the Urban Economy of Poland's Textile Metropolis in the 1990s." *Urban Studies* 30 (6): 1065–80.

Walsh, Maurice. 2007. "Zambia and copper." Taxing Questions, Part 2. BBC News, United Kingdom. http://news.bbc.co.uk/1/hi/programmes/documentary_archive/7091263.stm.

Ward, Kevin. 2003. "Entrepreneurial Urbanism, State Restructuring and Civilizing 'New' East Manchester." *Area* 35 (2): 116–27.

Way to Russia. 2006. "Ivanovo Guide." Accessed 2006. http://www.waytorussia.net/GoldenRing/Ivanovo/Guide.html.

Westwood, Sallie, and J. Williams. 1997. *Imagining Cities: Scripts, Signs, Memory*. New York: Routledge.

Whitten, Nick. 2006a. "Celebrations as Yard Wins Shipbreaking Green Light." *Evening Chronicle*, Newcastle upon Tyne, May 20. http://icnewcastle.icnetwork.co.uk/eveningchronicle/eveningchronicle.

Whitten, Nick. 2006b. "Poles Stage Wildcat Strike." *Evening Chronicle*, Newcastle upon Tyne, May 18. http://icnewcastle.icnetwork.co.uk/eveningchronicle/eveningchronicle.

Williams, Paul Harvey. 2007. *Memorial Museums: The Global Rush to Commemorate Atrocities*. Oxford: Berg.

Wilson, Janelle L. 2005. *Nostalgia: Sanctuary of Meaning*. Lewisburg: Bucknell University Press.

Winson, Anthony, and Belinda Leach. 2002. *Contingent Work, Disrupted Lives: Labour and Community in the New Rural Economy*. Toronto: University of Toronto Press.

Wylie, John. 2007. *Landscape, Key Ideas in Geography*. London: Routledge.

Yefimov, V.A., et al. 2006. *Ivanovskaya Oblast'*. Ivanovo: Id 'Referent.'

Yin, Robert. 1994. *Case Study Research: Design and Methods*. London: Sage.

Young, Michael Dunlop, and Peter Willmott. 1957. *Family and Kinship in East London*. London: Routledge & Kegan Paul.

Zavitz, Sherman. 2003. *It Happened at Niagara*. Niagara Falls: Lundy's Lane Historical Society.

Zukin, Sharon. 1982. *Loft Living: Culture and Capital in Urban Change*. Baltimore: Johns Hopkins University Press.

Zukin, Sharon. 1991. *Landscapes of Power: From Detroit to Disney World*. Berkeley: University of California Press.

Zukin, Sharon, et al. 1998. "From Coney Island to Las Vegas in the Urban Imaginary: Discursive Practices of Growth and Decline." *Urban Affairs Review* 33 (5): 627–54.

Index